IN THE OKEFENOKEE.

IN THE
OKEFENOKEE

A Story of War Time
and the
Great Georgia Swamp

By
LOUIS PENDLETON

The Black Heritage Library Collection

BOOKS FOR LIBRARIES PRESS
FREEPORT, NEW YORK
1972

First Published 1895
Reprinted 1972

Library of Congress Cataloging in Publication Data

Pendleton, Louis Beauregard, 1861-1939.
    In the Okefenokee; a story of the war time and the
great Georgia swamp.

    (The Black heritage library collection)
    SUMMARY: Two boys are captured by Confederate desert-
ers during the Civil War and held prisoner in the
Okefenokee Swamp.
    [1. U. S.--History--Civil War--Fiction.
2. Okefenokee Swamp--Fiction] I. Title. II. Series.
PZ7.P384Ino 7   [Fic]   72-1558
ISBN 0-8369-9045-5

PRINTED IN THE UNITED STATES OF AMERICA

IN THE OKEFENOKEE.

A STORY OF WAR TIME AND THE GREAT GEORGIA SWAMP.

BY

LOUIS PENDLETON.

AUTHOR OF

"KING TOM AND THE RUNAWAYS," "THE SONS OF HAM," "IN THE WIRE-GRASS," "THE WEDDING GARMENT," ETC.

BOSTON:
ROBERTS BROTHERS.
1895.

University Press:
John Wilson and Son, Cambridge, U.S.A.

CONTENTS.

IN THE OKEFENOKEE.

CHAPTER I.

SAD NEWS FOR REFUGEES.

IT was late in February of the year 1865, — winter according to the calendar; but already wild violets were peeping through the frost-browned wire-grass, and honeysuckle and dogwood blossoms had begun to perfume the air. In southeastern Georgia, winter is only a make-believe, and soon yields to spring.

Among the scattering pines in front of a "double-pen" log-house, and near a "wet-weather" spring, two boys were engaged in cleaning a gun; that is to say, one of them, the larger of the two, was thus occupied, while the other looked on with absorbed attention. The taller boy, who was about fourteen years old and well grown for his age, had removed his coat, and from time to time, as he paused in his work, wiped the sweat from his forehead. The other, who could not have been more than ten years old, stood where the sunshine fell full upon him, but had not yet found his coat too warm.

Both boys were dark and fine of feature, but hardly to be called handsome, although there was a straightforward, open look on the face of each which seemed to promise a manhood of truthfulness and honor.

The double barrels of the gun had been separated from the stock, and were held upright in a shallow tin basin of water from the spring. The ramrod, wrapped carefully with cloth, was drawn back and forth in the barrels, piston fashion, causing the water to be sucked in and sprayed forth from the tubes, and thus removing the accumulations of burnt powder and wadding.

"I'm goin' to give her a good cleaning this time, Charley," said the elder boy, "and maybe the next time I jump a deer she won't fail me."

"Papa says you don't clean your gun often enough," rejoined Charley, after a moment.

The larger boy appeared to disdain a response to this criticism at second hand; and there followed a long pause, during which fresh water was vigorously drawn in and sprayed out. Finally the younger boy spoke again, —

"Do you want to know a secret, Joe?"

"What is it?"

"I think Sister Marian must be goin' to marry Captain Marshall."

"Who said so?"

"Nobody. But I saw him kiss her."

"When?"

"Just before he left. He kissed her, and she did n't do a thing, — she just turned red."

"Humph!" was Joe's displeased exclamation. "I'm glad *I* was n't there. It would 'a' made me mad."

Little did the boys suspect that the young captain had obtained leave of absence, and travelled hundreds of miles, in order to ask for and obtain the right to do just what Joe so emphatically disapproved.

"And when he was gone, she cried," continued Charley. "I saw her."

"It was high time for him to go," said Joe. "Father says we need every man at the front we can get. He says the Confederacy is bleeding at every pore. It is such a pity that father is too old, and I am too young! I wish they 'd let me go, anyhow."

"Papa is over sixty," remarked Charley.

"Everything is goin' against our side," continued Joe. "Father says the Confederacy is 'tottering on its last foundations.' And to think that now, when every man is needed, the Okefenokee is full of deserters! Father said it made his blood boil. Brother George and brother Tom have been at the front from the very start," the boy added, with pride.

Joe had just finished cleaning the gun when Charley's attention was attracted to two men who were approaching the log-house from the woods. Half an hour before, he had seen his father go out with an axe to fell a tree. He was now returning, the axe thrown across his shoulder, accompanied by a neighbor, who held a newspaper in his hand.

What riveted Charley's attention was the fact that his father was weeping aloud.

"O Joe, look yonder! Papa must have hurt himself with the axe."

"You little goose!" cried Joe, turning to look. A moment later the elder boy's face changed, and with quickening breath he half-whispered, "Somebody must be dead!"

Then both boys started toward the house at a run.

That Mr. Roger Mérimée, the father of the two boys, was not of Anglo-Saxon descent might be surmised from the fact that he could weep in this way. He and his family did not belong to the Okefenokee backwoods. He was, or had been before the war, a wealthy rice-planter of the coast, living on the same spot where his ancestors, belonging to the persecuted and self-exiled Huguenots of France, had settled generations before.

Stern adversity was now the portion of this family; their beloved island home had fallen into the hands of a Union force, and they were refugees. Mr. Mérimée had brought his wife, daughter, and two younger sons, a few belongings and three servants, in boats of his own, up the St. Mary's River to the backwoods village of Trader's Hill, on the borders of the great Okefenokee Swamp. A mile from that settlement they had hastily erected a "double-pen" log-house. Here the family had sojourned — or "camped," as they said — during the past eight months.

The neighbor with the newspaper in his hand had not gone in. He halted at the gate a few moments, then turned to go. Observing the two boys running toward the house, he stopped, as if intending to speak to them, but after a moment's reflection moved on again.

The boys did not need to be told that a great grief had come to their home. As they came near the door, the sound of weeping issued from the large room on the right. Charley ran in, but Joe hesitated. Seeing a negro woman approaching from the kitchen, he ran to meet her.

"What is it, Aunt Martha ?" he asked, trembling.

"Mas' George —" she said. The woman's round good-humored face was now very sad. "Mas' George — " she repeated, falteringly

"Is he dead ?"

"Yes, honey."

The woman passed on hurriedly ; and Joe, after a moment, absently seated himself on a bench in the wide hallway, where from time to time the sound of fresh sobs reached him. The boy was only ten years old when his brother went to the war, and during the four years since they had seen each other but twice. They were almost strangers ; and it was only natural that Joe could not grieve as his parents grieved.

Still he was very unhappy. There was something appalling in this great grief which he could not fully share ; it filled him with anxiety and dread. He did not want to see them weeping, — his father, his mother, his sister ; it was painful even to think of. And so he stayed where he was, and waited.

While he waited, his thoughts were busy. He wondered where his brother was now, — his real brother, who would live forever, not the body which would be buried in the ground. Was he walking about in that world to which he had gone, and looking at things and asking questions; and were the angels teaching him, telling him everything he wanted to know? Joe thought there must be a great deal to see in that world, and that, but for the dying, — which every one seemed to regard as so very painful, — it must be very pleasant to go there.

Finally his sister Marian crossed the hall, and, observing him, approached. She was unusually handsome, in spite of her swollen eyes and tear-stained face.

"Have they told you, Joe?" she asked softly.

"Yes. Where was he when he fell?"

"At Columbia."

She burst out crying again, putting one arm round the boy's neck. In a few moments he, too, was overcome. The grief of the household had become his also. The future world might indeed be the delightful place he had pictured it, but the lifelong parting was terribly sad.

The house was astir at daylight next morning. Martha served as tempting a breakfast as the resources of the house would permit; but no one was hungry. Joe observed that his father, his mother, and his sister had each dressed with particular care that morning; and he wondered if they were going away. Presently his mother called him into her room.

"We are going down the river to St. Mary's," she told him, gravely. "We want to hear more news. Your father is too feeble to travel alone, and I must go with him. Marian will go, too; she cannot remain here without me, at her age. So we shall have to leave you boys with Martha and John."

Joe wondered why it was considered safe for Charley and himself to remain, and not safe for his sister; but he did not ask questions.

"Your father is not afraid to leave you with Martha and John," his mother continued. "They will be kind to you, and you must not do anything to provoke them, Joe. Of all our servants, they were always the best. We are not afraid they will run away and leave you, as Asa did." Asa was a negro who had belonged to the family and who had disappeared some time before.

A few minutes later the boys watched the wagon drive away. Martha and John were watching, too. It was a strange sight this, — their master and mistress seated in chairs in an open wagon and driving away, just as the commonest "Crackers" might have done.

The vehicle out of sight, John turned away silently to pursue the work which had been left him to do; but before Martha resumed her labors she said to the boys, —

"Yo' ma say you-all kin hunt much as you please while she gone, but you mus' be keerful. You-all better keep out dat swamp," she added, on her own account. "No tellin' what dem 'zerters might do ef dey cotch you in dat place."

Joe smiled contemptuously. What was a deserter but a cowardly sneak? And who was afraid?

The great Okefenokee Swamp, a wild waste some forty miles long by twenty-five broad, surrounded by vast tracts of pine-barrens almost without a settlement, is better known now than it was in those days, but its character is essentially the same. It consists now, as it did then, of vast jungles, flooded forests, islands, lakes, "prairies," or marshes, and is still comparatively a pathless wilderness.

More than a hundred years ago a story was current that it had been the last refuge of the ancient Yemassees, — a race which disappeared before the march of the conquering Creeks, — and it is well known to have been a stronghold of the Seminoles during the Florida-Indian war, as well as to have furnished a secure hiding-place for deserters during the Civil War. At present its more accessible islands sustain one or two squatter families, while the swamp itself is, as it has ever been, the bountiful and protecting mother of a variety of wild animals, birds, alligators, and other reptiles.

Joe and Charley had never ventured far into it, but had often, alone or with their father, hunted along its borders, and had, therefore, some idea of its general character. The elder boy was not lacking in courage, but was restrained by prudence. To say nothing of the possible encounters with reptiles, bears, and panthers, he knew that there were thorny jungles through which it was difficult to go without paying a penalty of torn clothing and bleeding limbs, and that there were vast marshes, wherein one often sank to the armpits in mud and water.

None the less, however, was there an alluring attraction about the great swamp; its remote recesses rose before the boy's imagination, unveiling their wonders and inviting his approach.

Joe had long been determined to extend his explorations when a favorable opportunity should arrive. The day after the departure of his parents he decided that the time had come. Permission to enjoy unlimited hunting had been given him; why not penetrate the Okefenokee, to the extent of two or three days' journey at least?

The chief obstacle in the way was Charley. Joe felt that the boy was too young to go, and yet he did not like to leave him behind. Nor could he think of going alone without misgivings. If he only had a comrade, a boy friend of his own age — or even if John, the black man, would agree to go. This, however, was out of the question; John had work to do, and in any case probably could not be persuaded to go.

But Joe felt that something must be done. He was not disposed to idle about the house, and dwell upon the grief which had befallen the family. If he could but find some of those deserters hiding in the swamp and tell them how things were going at the front, they — perhaps they would become ashamed of their evil way and return to their duty. Could he but accomplish this, how happy he would be! For hours the boy could think of nothing but this glorious plan.

However, he concluded to wait still another day before starting, hoping some one at Trader's Hill could be persuaded to go with him.

One plan after another suggested itself to Joe that afternoon, as he and Charley walked out to try the newly cleaned gun. Martha had given them an early dinner, and they had a long afternoon before them. Heedless of her repeated warning, they at the outset turned their steps in the direction of the great swamp. This was but natural, for there was less game in the pine-barrens.

Joe trudged ahead, his gun across his right shoulder, and a powder-horn and shot-pouch hanging at his left side. Charley followed, armed only with a hatchet; he was considered too young to handle a gun.

For about two miles the path led through open pine-barrens, carpeted with wire-grass, level as a floor; then gradually a downward slope was perceived, and ere long the straggling pines were merged in the thicker growth of the swamp.

Quitting the path which skirted the swamp, Joe led the way through a "head," or arm of the great morass, thickly grown up with cypresses and covered for the most part with shallow water, through which the boys boldly waded. It did not occur to them to remove their shoes, or to take a circuitous route in order to avoid the water. To penetrate the Okefenokee even for half a mile with dry feet was out of the question. An hour later, after following a dimly out-lined trail for some two miles, the boys found themselves on the shore of a little lake or pond, the surface of which, except near the centre, was largely hidden by "bonnets" — a species of water-lily — and clumps of brown flags or sedge.

Charley had never been so far before, but Joe remembered hunting along this lake with his father, who had shot three ducks. The deserters were now forgotten; and visions of wild ducks, both alive and slain, floated before Joe's inner sight and urged him on.

He skirted more than half the way round the lake, creeping forward stealthily, before he sighted a flock of ducks within range. Then he was so much excited that his aim was wild and fruitless. Charley, who had been directed to remain quiet and far in the rear, now hurried up to see what Joe had shot.

The sun was fast sinking behind the wall of woods; and Charley insisted that they should at once turn back, or night would overtake them. But Joe refused to turn back until he had skirted the lake twice, shot several times, and finally killed a duck, to secure which he waded up to his waist in the sedge.

Struggling out of the water with his prize, the boy hurriedly took his bearings and led the way along what appeared to be the trail by which they had come.

Within an hour the sun had set and the twilight was thickening. This would have mattered little if they had been clear of the swamp; but so far from having gained the open pine-barrens, they now seemed more deeply involved than ever, and were unable to recognize anything about them.

Joe halted and looked anxiously around. He suspected that, in skirting the lake, intent on the game only, he had lost his bearings, and in starting homeward they had taken the wrong direction. This, indeed, was true.

"Don't be frightened, Charley," he said manfully, after a few moments; "but we are lost, and we shall have to stay here all night!"

CHAPTER II.

LOST IN THE OKEFENOKEE.

"STAY here all night!" cried Charley, gazing around the gloomy swamp through starting tears. "I *said* we ought to turn back before."

"Yes, it was all my fault," said Joe; "but it can't be helped now."

"Do you think the panthers will smell us and — and — come?" asked Charley, in a whisper.

"Don't be foolish. We are n't far enough in for that," answered Joe, stoutly, although the last part of his speech sounded a little weak, as if he had misgivings. He had never spent a night in the swamp; and the prospect of it now, under the existing circumstances, was little short of terrifying. But he said resolutely, "It's no use to think of finding our way home to-night, and we had better hunt a place to camp right away."

Promptness was indeed necessary, for it was fast growing dark. After a hurried search the boys selected a little open spot which was comparatively dry, and covered with dead grass. Within two or three feet stood a large black-gum tree, which, Joe reflected, could be climbed easily in an emergency; and close at hand was abundance of hemleaf and huckleberry bushes. The tops of these could be broken and piled where the boys expected to sleep, and the couch thus prepared, though not likely to suggest down, would at least protect them from the damp ground.

Joe began next to collect fuel, as he should have done at first. They had scarcely begun to do this when it became so dark that no object more than three feet distant could be distinctly seen. Dry wood appeared to be very scarce. They had not as yet secured even a good torch, and Joe wasted more than half the few old and broken matches found in his pockets in an anxious search for a piece of "lightwood."

Even then he did not find what he wanted, and began to consider giving up the fire. It certainly would not do to be left without a match. Who could tell when they would find their way out of the swamp? Perhaps, after all, it might be better to pass the night without a fire, unless they could have a very large one. A small blaze could hardly frighten, and might attract wild animals.

Joe struck one more match with no better result, and then gave up in despair. They now applied themselves to breaking and heaping the brush, and presently lay down upon the pile.

Although in the swamp the darkness was dense, it was a clear night, and an occasional star could be seen through the foliage. After silently reciting their prayers, the boys lay close together, occasionally speaking in whispers and looking wearily up at the stars. At every sound in the forest, at every freshening of the night breeze in the leaves, they would start and listen, apprehending the attack of some wild animal.

Although the month was February, it was a balmy spring night. But the boys were without covering; their feet and legs were wet, and they soon began to feel cold.

Presently Joe rose and broke more of the huckleberry tops; making Charley rise, too, he scooped out a hollow in the enlarged pile. Then they lay down within it, covered themselves up to their ears, and felt warmer.

Nothing disturbed them for a long while except an owl which lighted in the black-gum, and repeatedly demanded to know, " *Who-who-who-all ?* " as Charley declared. But after it flew to a distant perch, all was quiet except for the occasional rustling of the branches, and at last the weary boys fell asleep.

Some hours later Joe was awakened by feeling Charley move, and hearing his voice close to his ear, —

" Joe, Joe, wake up ! I heard something ! "

Joe was wide-awake in a moment. Listening intently, he heard a stealthy footfall, then another and another, circling round the camp. The sounds could hardly have come from more than thirty feet away.

" Let 's climb that tree ! " proposed Charley, excitedly. " It may be a panther ! "

A twig snapped under the foot of the prowling animal, and terror seized the boys. Grasping his gun and ammunition, Joe leaped to his feet and bounded to the tree, Charley close at his heels. Every moment they expected a panther to spring.

Joe held back, and let Charley go up the tree first, helping him until he could grasp the lower branches. Then, having passed up his gun, the elder boy climbed nimbly into the tree. Lodged in the branches of the black-gum some twenty feet from the ground, they listened intently, but heard no further sound. The marauder appeared to have been frightened in turn, and had either retreated, or had squatted and was remaining quiet.

An hour passed, and still there was no sign. Arranging themselves as comfortably as possible among the spreading branches near the tree's main stem, the boys began to forget their situation and to doze.

Awakening with a start some time later, Joe caught a glimpse of two gleaming eyes beneath the tree. Making sure of his gun, he whispered to Charley, who also began to stir, —

"Do you see him ? Do you see his eyes ?"

But Joe had scarcely opened his mouth, when a low, guttural growl advised him that he had seen aright. Raising his gun, he tremblingly pointed it downward, and as soon as he saw the eyes again, aimed at them hastily and fired.

The gun's report was followed by a howl of pain ; and then, during some moments, they could hear the wounded animal beating a frantic retreat through the neighboring underbrush. The boys were well satisfied to find that the scattering duck shot, even if they did not kill, would wound and drive away this panther, bear, wild-cat, or whatever it was. Joe remarked cheerfully that it was a great thing to have a gun, and both boys felt more comfortable after this, although they dared not descend from the tree.

An hour later day began to break ; but the obscurity still shrouding neighboring objects for some time thereafter was entirely dissipated, and the sun was well up before the boys left their perch. Meanwhile Joe outlined plans for the day.

"Charley," said he, "we'll go back on our tracks to the lake, go all around it carefully, make sure of the right path, and start off toward home. If we have good luck, we'll get there by dinner-time."

As they descended from the tree, Charley espied the hatchet near their bed of leafy boughs, and picked it up. They then observed that the ground was covered with feathers, with here and there a few fragments of small bones, and recollected the duck which Joe had shot. Evidently the animal which had visited them in the night had enjoyed a feast at their expense.

"It may have been only a mink," said Joe, almost disposed to laugh. But he added, "I think it must have been at least a wild-cat, though."

"It scared us just as much, anyhow," said Charley.

Full of hope, they cheerfully started off on the backward trail. For the first half-mile it led over soft boggy earth, where the tracks were easily seen; but by and by they reached a tract of several acres dotted with clumps of palmetto-bushes, where the ground was firm and thickly covered with wire-grass.

Here the trail was soon lost. After some time spent in a vain attempt to find it, they pushed forward in what seemed the right general direction, hoping to pick up the trail. About an hour later they espied a sheet of water ahead of them.

"There's the lake!" they shouted together. But on

reaching its shores they found that it was not the lake wherein the duck had been shot, but another very much like it.

It was now plain enough that they were seriously lost, being several miles within the border line of the Oke-fenokee, and ignorant which way to turn. They looked about them in despair. Poor Joe had long since forgotten his great plan of seeking out the deserters, and now thought only of finding the way home.

He was not so disheartened, however, as to neglect a chance which offered for a shot at some ducks, and was for a few minutes highly elated on discovering that he had killed two, and that they were within reach. It was now near noon, and both boys were ravenously hungry.

They soon halted, therefore, at a little stream which ran into the marshy lake, built a fire, and prepared one of the ducks for food. The novel experiment of cutting thin slices from the bird, suspending them from the points of long sticks, and holding them close to the flames, absorbed their attention for a long while. Although the flesh of the duck thus roasted satisfied their hunger, and they considered it a very fine dish, they would under ordinary circumstances have regarded it as unpalatable in the extreme, owing to the lack of salt.

" The thing for us to do, Charley," said Joe, as they rose, a little more cheerful, to move on, " is to keep pushing ahead where the swamp seems open. Maybe we 'll find our way out after awhile."

They pressed forward on in this way for several miles during the afternoon, but at sundown their prospects did

not seem to have improved. They knew no better than before where they were. As it was clearly necessary to
remain in the swamp another night, they halted in time to
select a favorable spot for a camp and collect a large pile
of firewood.

Having cooked and eaten the second duck, which
Charley, with forethought, had brought along, drinking as
much of the swamp water as they dared, they built a
second fire some twenty feet from the first. Arranging
midway between the two a bed of collected moss, leaves,
and grass, they passed a quiet and fairly comfortable night,
without alarms.

The next morning they made an early start, and pushed
bravely forward, after making a poor breakfast by picking
the bones of the duck. Toward noon they were confronted by a seemingly impenetrable jungle.

"We'll have to turn back now," said Charley, dole-
fully.

"No, let's go right ahead," proposed Joe. "We'll have
to travel slowly; but I know we can get through it, and
maybe when we *do* get through, we'll be out of the swamp.
I've seen just such places on the edge of the Okefenokee
from the outside. I think the swamp has a thick rim just
like this round a great deal of it."

"Let's get some fat lightwood splinters for kindling-
wood," said Charley, "because we may be in that thick
place all night, and can't start a fire. It's low and wet
down in there."

This prudent suggestion was acted upon. They found
some good lightwood; and Charley carried the bundle of

splinters in addition to the hatchet, as Joe led the way with the gun.

The jungle evidently covered thousands of acres, and was for the most part so dense as to be penetrable only where wild animals had made their trails. The larger forest trees were not altogether absent here; but the jungle consisted chiefly of smaller trees, shrubs, and vines.

Among these was the " bamboo brier," a vine sometimes an inch thick, armed with thorns which pierce like knives, and the tangled growth of which occasionally forms an impassable wall ten feet in height. Besides all this, the ground was wet and boggy, for the most part indeed covered with water varying from two inches to two feet deep. It was not a great while before they bitterly regretted their decision to force their way through this jungle.

Often they had to bring the hatchet into use before they could move forward even a step; and their progress was so slow that, from about eleven o'clock in the forenoon until sundown, they pushed forward hardly more than two miles. As the sun declined, they were prey to growing uneasiness, but still pressed on. The hope that the terminus of the jungle was not far ahead led them forward; and indeed it was now idle to turn back, as night would arrive long ere they could retrace their steps.

Aware that little more than half an hour of daylight was left them, the boys halted at a point where the jungle was somewhat less dense than usual in order to make some preparations for the night. But even here the water rose above their ankles, and the prospect was a very gloomy one.

They had often heard how belated Okefenokee hunters had been compelled to build sleeping bowers whereon to pass the night, and this they set about doing without delay. Selecting two saplings about eight feet apart, they cut into them with the hatchet at a point about three feet above the water, until they toppled and fell over in the same direction. These saplings, being young and green, did not entirely separate from their stumps; and therefore, while slanting gradually down to the water, offered a support to the smaller poles and brush with which the boys bridged across from one to the other. The resting-place thus secured was extremely uncomfortable, but was better than spending the night in a tree, — the only other recourse open to them.

It was now dark, and they attempted to build a fire in the hollow of a cypress "knee" within a foot or two of their sleeping-bower. But they were unable to gather together sufficient dry fuel; and, wisely determining to reserve some of their lightwood splinters for an emergency, the little flame was presently allowed to die out, leaving them in deeper darkness than before.

As they rested there, scarcely daring to speak above a whisper, they were thankful for one thing, — that it was yet too early in spring for moccasins and other reptiles to be abroad. This thought was only as a bright ray in the gloom, however.

Lying on an uncomfortable pile of boughs three feet above the stagnant water, in hunger and darkness, without the hope of finding the way home, their distress of mind and body was very severe. Charley broke down at last, and sobbed himself to sleep.

Joe made a manful effort to say comforting words, reminding his small brother how often their father had told them that all things were for the best in some way; and that the Divine Providence never forgot them. But it was difficult to take comfort from these reflections at such a time, and Joe himself was painfully depressed. Fatigue overcame him, however, and by the time Charley's sobs were stilled, he, too, was asleep.

If there was any tramping of wild animals about their camp that night, the boys did not hear it. At an early hour of the morning they were awake and preparing to push forward, although very far from having recovered either from the mental or physical depression of the previous night.

About nine o'clock, to their great delight, they emerged from the jungle and ascended the slope of an open pine ridge, upon which, at a distance of some three or four hundred yards apart, they noted three Indian mounds about fifteen feet in height.

Joe now believed that they were out of the swamp; but a two-hours' tramp was sufficient to convince him that they were merely on an island about three miles long by one mile in breadth, and that they were probably farther away from help than ever.

In the course of their tramp Joe had shot two partridges, and the two lost boys were in a measure comforted by the thought that they at worst need not starve; and presently they made a discovery which brought fresh hope. At the farther end of the island, where a dense "hammock" sloped down and joined hands with the swamp, which here took the form of a flooded forest, they found a boat, — a small bateau scarcely capable of floating three persons. Evidently it had been lying idle for some time. It was half-full of water; but when this was bailed out, it showed no serious leaks, and carried the two boys safely.

"That must lead out to a lake," said Joe, indicating the narrow boat-road which could be clearly seen winding away through the flooded forest. "And once on that lake, we may find our way out of the swamp! Anyhow, we may meet somebody."

Halting only to build a fire and broil and eat the partridges, they got aboard the boat with all their belongings, and paddled away. The boat-road had evidently been a good deal travelled, and it was not very difficult to make headway. As Joe had surmised, it led after a few hundred yards into a lake, — a long narrow sheet of water which was in reality a "dead" river. At its farther end the boat-road began again, and wound on its way as before through the seemingly endless flooded forest.

Along here the boys suddenly caught sight of a large animal swimming across their path some fifty yards ahead. Gazing at it in breathless astonishment, they quite forgot the gun until it was too late to shoot.

Charley feared it was a panther, but Joe said it was probably only a wild-cat. As they neared the spot, he stood up, gun in hand; but the hurrying beast had landed in the jungle, and no sign of it could be seen.

A mile or two farther on, they emerged from the flooded swamp upon an extensive open marsh filled with long rushes and "bonnets," and dotted with small islands and clumps of trees, hung with long gray drifts of Spanish moss.   As far as the eye could reach, straight ahead, to the right or to the left, nothing else was to be seen.

Here the boys paddled for hours, imagining that they were pursuing the same general course, but in reality wandering widely in the confusion of rounding many little islands.

At last they saw far ahead the tops of some tall pines, and gradually worked their way toward them, surmising that they stood either upon a large island or the mainland.

As they approached within half a mile, a shallow marsh, free of clumps of trees or little islands, opened before them.   In the shallower water here, the rushes and water-mosses seemed to thicken steadily as they neared the shore, and it became more and more difficult to force the bateau through or over them, although the boys followed the windings of a clearly defined boat-trail.

Finally, within some three hundred yards of the shore, or the wall of woods indicating an island, they were compelled to step out and drag the boat after them, sinking now to the knee, now to the waist, in slimy moss, mud, and water.

Entering the border of trees, they pushed forward, still in water knee-deep, for about a hundred yards, before they reached a landing-place where two boats, somewhat larger than their own, were moored.

"There's somebody here, *sure*," said Joe, looking about hopefully.

CHAPTER III.

THE DESERTERS' CAMP.

A WELL-BEATEN path led upward through the dense hammock between the swamp proper and the pine ridge composing the island upon which Joe and Charley had disembarked. As it was now near sundown, and the boys were painfully hungry, they did not pause to think twice, though they looked ahead warily as they followed up the path. The hammock growth here was largely of bay and magnolia, with a tall underbrush of swamp-cane. Emerging from this near the top of the slope, some two hundred yards from the boats, they found themselves in a small clearing, beyond which the open pine land of the island stretched away monotonously.

Near the centre of the clearing was a house, built of rough logs and puncheon boards, and elevated some twelve feet from the ground on stilt-like posts ; and over a fire to the right of this structure bent a man's figure. Evidently he was cooking his evening meal, for the boys caught the delicious odor of frying meat.

"Maybe he'll give us something to eat," said Charley, wistfully.

Just then the man stood erect ; and they saw that he was a negro, in a dirty homespun shirt and ragged pantaloons. A moment later he turned his face toward them.

"It's Asa!" said Joe, astonished.

The boys hesitated no longer. The negro heard their steps, and looked up. The bewildered expression which overspread his face changed quickly to one of delight. He leaped forward to meet them.

"Well, well, you boys!" he cried, laughing. "Where you-all come fum? Wut you doin' yuh?"

"What are *you* doing here?" asked Joe, halting at the fire.

But Charley broke in to outline in a few hurried words the story of their wanderings. He shared all the negro's delight in the meeting; but Joe, though glad enough, had not forgotten what he regarded as a very grave matter.

"What are you doing here?" he repeated, as soon as there was a break in the negro's exclamations. "What made you run away, Asa?"

"Me run away! Did you-all tink I run'd away?" asked Asa, an injured look overspreading his face. "De 'zerters cotch me an' brung me yuh — *I* never run'd away. No-suh-ree! One evenin' I was down in de edge o' de swamp huntin' yo' pa's cows, an' de 'zerters run out de bushes an' grab me an' tied me an' brung me in yuh, an' yuh dey been makin' me do dey cookin' an' all dey dirty work. Hit's de fac'. You des wait an' see now."

There was an air of sincerity about the negro which made the boys believe him. Besides, they remembered that he had always been a favorite in the family, and had never run away before.

His color was deep black, and his features were more pleasing than those of the average negro, and a certain intelligence and gravity of the eye inspired confidence. He looked quite young, but his age may have been anywhere between twenty-five and forty years.

"So this is the deserters' island," said Joe, glancing around. "How many live here?"

"Der's eight of 'em on dis islan', an' mo' on some de others."

"Where are they now?"

"Dey ain't come in yet. Some of 'em runnin' a deer, an' some gone ter de traps." Asa pointed to the skins hanging from grape-vines stretched beneath the house, and also beneath a low shelter of thatched palmetto fans. "Dey in de trappin' business," he added.

At this moment some one was heard coming through the bushes, singing in a peculiar childish voice, —

"Open the gates as high as the sky
And let King George's army pass by."

"Dat's Billy," said Asa. "He ain't got good sense, you see 'im so."

A barefoot young white man, clothed in rags, entered the clearing at a trot, and ran up to the two boys. Fixing his eyes on Joe, he inquired with a giggle, "What's your name?" When Joe had told him, he turned to Charley with the same question.

His hair was light in color and soft as a child's; but his face was as deeply wrinkled as many an old man's, and wore a curious, meaningless smile. His pale blue eyes were vacant, yet restless.

"*He* isn't a deserter, is he?" asked Joe of Asa, aside.

" No; but he belong to one. He's Sweet's nigger, an'
I'm Bubber's," said Asa, showing his white, even teeth.
" I waits on Bubber, an' Billy he waits on Sweet. Bubber
stole me, you know, so I'm his'n. I reckon Sweet stole
Billy, too; he had 'im yuh waitin' on 'im when I come."

" Who are they, — Bubber and Sweet ? "

" Mr. Bubber Hardy an' Mr. Sweet Jackson is de ring-
leaders o' de 'zerters," explained Asa.

In almost every Cracker family there is a "Bubber," —
a little boy whose brother or sister lisps out "bubber" in
trying to say brother. The nickname sometimes follows
an unfortunate boy to manhood. So had it been in the
case of "Bubber" Hardy, who, according to Asa, was
" cock of the walk" among the deserters. He was a great
stalwart fellow, with a waste of muscle and of a kindly
disposition.

Of hardly less importance was "Sweet" Jackson, —
another illustration of the tenacity of Cracker nursery
nicknames, — who was second only to Bubber in size,
muscle, and consequent authority. He was less popular,
however, being sullen and ill-tempered.

" When he git mad he don't no mo' mind knockin' Billy
aroun'," continued Asa, looking toward the half-witted boy,
who was still questioning Charley. " Bubber ginnerly
give me ter understan' I got ter be spry an' wait on him
right; but he don't never jump on me like Sweet do
Billy."

Further description of the leading deserters was now
cut short by the sound of approaching footsteps; and Asa
turned hurriedly to the fire, where he had been frying corn-

bread. The boys looked around in time to see a large man clad in dirty homespun advance from the borders of the darkening woods, a rifle over his arm, followed by two others carrying a small doe suspended from a stick which ran across their shoulders. Several dogs accompanied the party.

"Dat's Sweet," whispered Asa, as the leading hunter approached.

The two men threw the deer down on a carpet of palmetto fans, and immediately began to skin it, merely glancing once or twice at the boys. The leading hunter, who, according to Asa, was Sweet Jackson, presently showed more curiosity.

"Who-all's this ? " he cried gruffly, approaching the fire. "Billy, git me some water, quick ! Whar did you boys come from ? "

"From Trader's Hill, or very near there," answered Joe.

"An' what you doin' 'way h-yuh in the Okefenokee ? " he asked, adding, with a sudden suspicious gleam of the eye, "They sont you in to see whar the deserters was, did they ? They played thunder if they did."

"We went hunting in the edge of the swamp and got lost," answered Joe, simply.

"Well, an' how did you git across the perrarie ? "

The boys told him how they had struggled through the great marsh. The man asked several more questions, all indicating suspicion.

In the midst of Joe's explanation another party of hunters came out of the dark woods, exhibiting an otter

skin as their only but by no means insignificant trophy.
Among them was the "cock of the walk," Bubber Hardy.
Standing in the background long enough to hear the outline of the boys' story, he approached them in a more
friendly way than any one else had as yet done.

"How you come on, boys?" he said, extending his hand
to Joe. Then, turning to Charley, "This one's as putty
as a little gal," he continued, smiling admiringly. "He
outfavors his brother."

Charley was highly indignant at this; but both he and
Joe felt intuitively that the "cock of the walk" would
prove their best friend among the deserters. As he put a
few questions to them and listened to their straightforward
answers, they observed him narrowly.

He carried an army rifle, like the others, and was dressed
in homespun, the loose, ill-fitting fabric serving to give
him the appearance of being heavier than he really was.
He was above six feet tall, and evidently an uncommonly
muscular and powerful man. What attracted the boys
was the kindly gleam of his eye and an expression of quiet
resolution in his face, which was rather more handsome
and intelligent-looking than that of any of the others.
The boys wondered that such a man, who looked brave
if he was not, should have become a deserter.

Meanwhile Asa had been busy frying thin strips of the
fresh venison steak, and now announced that supper was
ready. The men silently took their places round the fire,
eating and drinking heartily.

The boys had not eaten since morning and were ravenously hungry, but did not move from their place, as no

invitation was given them. However, they were not neglected. At the bidding of Bubber, his master, Asa invited them to sit on the grass, placed a palmetto leaf between them, and piled it high with fried steak and bread. Later, he gave each of them a cup of "corn coffee."

The hapless Billy, who had taken the liberty of appeasing his hunger before the others began to eat, now lay on the ground, singing in an aimless, tuneless sort of way :

> " Meena— myna — mo —
>   Ketch a nigger by the toe.
>   If he hollers, let him go."

The young man's mind was evidently still in its childish state, and dwelt with delight on nursery rhymes. When Joe and Charley had satisfied their craving for food, and begun to observe him more closely, he was declaiming :

> "Queemo — quimo — dilmo — day
>   Rick — stick — pomididdle — Dido —
>   Sally broke the paddle over Mingo's head!"

He was beginning, "One-two, buckle my shoe — three-four, open the door — five-six, pick-up-sticks," etc., when Sweet called his name roughly, and sent him on an errand.

"What's the news about the war?" asked Bubber of Joe, as the men lighted their pipes and settled into comfortable lounging positions about the fire.

"Very bad," the boy answered, with a sudden trembling of the lip as he thought of his dead brother. "Everything is going against our side."

"I'm mighty sorry of it," rejoined Bubber, gazing into the fire abstractedly.

" Well, I ain't a-carin' so much," said Sweet. " ' T ain't
none o' my lookout. They kin settle it 'twixt 'em.''

Several of the men grunted approval at the close of this
speech. Nevertheless, Joe, who was becoming greatly excited, dared to bestow a look of contempt on the speaker..
Then, looking steadily at Bubber, he blurted out, —

" I don't see how you men can have the heart to stay
hid in here, when every single man is needed at the front.
I — I — I'd be *ashamed !* ''

Bubber winced. Sweet sat erect with a threatening look,
and some of the others uttered ejaculations of astonishment.
Still it was evident that the boy's boldness had excited
admiration. Joe, however, did not perceive his advantage,
and for the time his courage failed him. The pause was
broken by Sweet.

" Who 's ashamed ? " he cried with derision. " I ain't,
for one. What 's the use o' beatin' an' bangerin' aroun' ?
' T ain't none o' my quiltin'. I ain't got no niggers to
fight for."

This was too much for Joe. " What 's that got to do
with it ? " he cried indignantly, and began to speak excitedly of State's Rights and other features of the Cause,
in language borrowed from his father.

" It 's got a heap to do with it, I 'm a-thinkin'," Bubber
Hardy remarked, as the boy paused, conscious of his imprudence. " Them that don't own niggers, like me, naturally
ain't got the same interest in it. And yit I ain't proud o'
bein' a deserter — not a bit. But, niggers or no niggers, I
had good reasons. If anybody thinks I deserted jes' becaze
I was a-scared to fight, I jes' want him to stand up right
now and say so."

After this challenge there was a pause. Then Bubber began to talk about an occurrence in the day's hunting. By and by the conversation dragged. All were becoming drowsy. One by one the men rose and disappeared, until only Sweet, Bubber, and the two boys were left. Then Sweet rose and said to his comrade, —

"What you aim to do with them boys to-night, Bubber? We got to keep our eye on them boys."

"They 'll sleep with me," said Bubber.

Shortly after this, Hardy lighted a torch, and bade the boys follow him. He led them beneath the curious log-house standing so high in the air, — a precaution against snakes in summer, — and climbed by a ladder through a square opening in the floor.

Passing the sleeping men, whose hard, wrinkled faces seemed somewhat softened in slumber, Hardy led the way to the extreme end of the room, and, giving the torch to Joe, began to scatter and broaden his really comfortable bed of leaves and Spanish moss, so as to make room for the boys between himself and the wall.

Before the light was put out, Charley inquired where Asa slept, and was told that at night he was kept shut up in a little room at the opposite end of the long sleeping apartment. There was no window in all the structure, but enough air entered between the logs of the walls and through the door in the floor.

The boys were too weary to waste much time in worrying about their situation, and soon forgot everything in sound sleep.

CHAPTER IV.

PRISONERS.

WHEN Joe and Charley awoke next morning, they were alone in the sleeping-loft. Descending the ladder, they found Asa at the fire with something for them to eat; and after they had washed their hands and faces, Asa pouring water for them, they ate heartily. All but two or three of the deserters had gone off to the traps, or hunting, and these two or three were nowhere to be seen just now. By the time Joe and Charley had made a breakfast, however, Bubber appeared.

"Well, boys, what you aim to do?" he asked in a friendly way.

"I'll tell you what I'd *like* to do," said Joe, earnestly, encouraged by his tone, "and that is, persuade you, and as many of the rest as I could, to give up this—this deserting—and go back to the war again."

Bubber laughed outright. "I depend you've laid out to do a big job of work," said he; "most too big, I reckon. Better give it up. Better jes' stay h-yer a while with us, and learn to hunt."

"I would n't mind staying a while if—if there was a chance of persuading—"

"But ther' ain't, though, so you'd better not bother your head about it, son."

3

"Well, then, all I can do is to take Charley and Asa and go home."

Bubber laughed again, more heartily than before.

"I don't much think the other gents 'll be willin' to part with you and Charley yet a while. They loves comp'ny, you know! We all talked it over this mornin'. And as for the nigger — well, I don't see hardly how I could spare him."

"He's not your negro," cried Joe, indignantly. "He belongs to my father, and I'm goin' to take him, too."

"He b'longs to your father, shore enough," rejoined Bubber; "but, you see, I borryed him, and as they use to tell me, possession is nine points of the law."

Joe turned away angrily, and, calling to Asa, bade him make ready to start for home. He was too much excited to see how utterly powerless he was.

"I glory in your spunk, boy," remarked Bubber, quietly, "but I think you are wastin' it. If I was in your place, I'd know better than to be so rambunctious."

Joe made no reply, and repeated his order to the doubtful, hesitating negro.

"Listen to me," said Bubber, sharply. "If you walk off from h-yer with that nigger, it won't be five minutes before he'll be knocked down and dragged back, and you and Charley 'll be put under lock and key. I don't say *I'll* do it, but it'll be done!"

Joe now began to realize his position. Not merely was Asa a prisoner in the deserters' hands, but he and Charley as well. The latter could control himself no longer, and began crying.

"Look h-yer," said Bubber, "if we was to let you and that nigger go, fust thing we'd know you'd be guidin' a company of soldiers to this h-yer islant, and the last one of us would be led out and shot."

Joe was conscious of a strong impulse to bind himself by a solemn promise against any such action, but checked it as weak and unworthy, as he thought of all that was involved.

"If you'll agree to leave the nigger and say nothin' to nobody when you git home," continued Bubber, as if divining the boy's thoughts, "maybe after a while I kin persuade the boys to let me take you across the perrarie and put you on the trail to Trader's Hill."

"I won't agree," said Joe, stoutly, although tears started in his eyes, and Charley's sobs were louder than before.

"All right. You'll stay right h-yer, then!"

So ended their conference.

"Never mind, Charley; don't cry," said Joe, bravely, as soon as Bubber was out of hearing. "We'll just have to watch our chance and make our escape, that's all. Have you tried to escape yet, Asa?"

Asa answered with a grunt that he had tried it once. He had gone one day with three of the deserters in two boats to the country across the "prairie" in order to cut a bee-tree, and while there had made a dash for liberty; but he was soon caught, and the whipping he had received was a warning not easily forgotten. He had never tried it again.

"Well, we must watch our chance," Joe repeated.

But before the day was gone he realized that the op-

portunities likely to occur would be few and far between.
The boys were free to walk about the camp, but were
always under watch.   While the rest of the men were
away hunting and trapping, at least two were always in
sight, either inspecting their stock of hides, or lounging
about lazily, drinking corn-beer of their own brewing, and
telling yarns.   Asa was also free to come and go within
certain bounds; but when he was not engaged in bringing
wood and water, cooking the meals, or waiting on Bubber,
he generally lay tamely on the grass in the sun and dozed.

A certain sympathy and friendship existed between him
and the half-witted Billy.   They were fellows in misfortune.   But after the coming of Joe and Charley the
hapless youth transferred his attention to them.   Charley
particularly seemed to please Billy.   He hung about the
camp during all of that first day, talking sense and non-
sense alternately, and repeating many nursery rhymes.

"I like you," he said to Charley once.   "Some o' these
days I'm goin' to take you to see son."

"You haven't a son!" said Charley, laughing.

"Wait till I show him to you, and you'll see."

"Who is he?"

"Never you mind," answered Billy, almost exploding
with mirth.   "You'll find out some day; you'll find out,
boy.   I must go and see son now," he added later, with
his strange laugh, and walked off into the woods.

All the deserters but Bubber and Sweet went away
early the next morning, — some to hunt, others to visit the
many traps which had been set here and there on the
island and in the surrounding swamp.

Asa had just finished his labors after breakfast, and Bubber was lounging near, talking amicably with Joe and Charley about hunting, when Sweet walked up and asked :

" You goin' to use Asy this mornin', Bubber ? "

" Not partic'lar."

" Well, I'd like to borry him. I'm goin' to build me a permeter shelter for my own hides, so I kin spread 'em out more."

" All right."

Thereupon Asa, who, it would appear, might be " used " and " borrowed " like any inanimate thing, was led away in company with Billy. Their business was now to cut down one six-inch sapling for posts, and several two-inch ones wherewith to frame the slanting roof which these posts would support. This done, they must gather hundreds of palmetto fans and thatch the roof, all under the direction of the ill-tempered Sweet.

The three had been thus engaged some thirty minutes when Bubber, Joe, and Charley, at the camp, heard sounds of blows and screams. A few steps toward the spot selected for the palmetto shelter revealed the cause of the uproar.

Sweet, completely out of patience with the half-witted and trifling Billy, had fallen upon him, and was whipping him with a long supple stick. As he laid on his blows more and more fiercely, in spite of his victim's piteous cries, the boys drew near in horror, slowly followed by Bubber.

" Stop that ! " cried Joe, hotly, as he arrived on the scene.

"I 'll stop when I git ready!" retorted Sweet, in a fury, pausing for a moment. "And if you give me any yo' sass, I depend I 'll wallop you in the bargain. You 're 'most too spargy for me, anyhow. You 're gittin' too big for yo' breeches."

"You coward!" cried Joe, as the blows recommenced. "You ought to be ashamed to beat that poor half-witted —"

Here Sweet suddenly let Billy go, and turned upon Joe with uplifted stick.

"Hit him if you dare!" said Bubber, stepping up to them.

"'T ain't none o' yo' business, Bubber Hardy!" cried Sweet, threateningly, turning to meet the new attack.

"Hit 's everybody's business when you jump on that poor boy Billy that way. You know he ain't accountable."

"I reckon I 've got a right to thrash him if he won't work! I kin hardly make him lift his hand to do a thing, and when he does work he works so powerful sorry —"

"I thought you was more of a man, Sweet Jackson."

"I depend I 'm man enough to give you all you want!" the man replied with an oath, making a threatening movement.

Bubber caught one end of the uplifted stick; it broke between them, and they closed in hand-to-hand combat. Luckily, neither was armed; if either had been, bloodshed must have followed. As it was, they were well matched, and it was evident that the fight must be a long one.

Joe was too much absorbed in the conflict to see the

opportunity which it offered; but Asa, less excited by such a scene, thought more quickly.

"Now de time!" whispered the negro, in a low, cautious voice over Joe's shoulder. "Less slip off an' run down to de boats. Ef we git dem boats, we kin git away. You an' Charley kin take one, an' Billy one, an' me one. Ef we git out on dat prairie 'mongst dem islants, we out o' dey reach. Dey can't come atter us far widout a boat."

The negro began to move away, calling softly, "Come on, Charley!" and beckoning in a commanding way to Billy. Neither Charley nor Billy understood what he meant, but both were attracted by his mysterious manner, and followed him.

Joe hesitated, his glance returning to the two combatants. He wondered if it were quite honorable to sneak away while Bubber was fighting in his cause as well as Billy's. Still, he and Charley and Asa and Billy were unjustly held prisoners, and if there was a real chance of escape, why not go?

The boy thought of his parents, of his sister, of his dead soldier brother, of the cowardly men who had deserted in the hour of direst need, — after all, the kindly Bubber was only one of these. This decided Joe. The boy saw that Asa was now as far as the camp, and Billy and Charley were close behind him. Charley caught Joe's eye, and beckoned. Slipping behind a clump of bushes, Joe ran to the spot where his gun stood.

Passing the camp, Asa caught up a tin bucket of sliced venison and an axe, then darted along the winding path through the swamp cane toward the boat landing. As Joe hurried along the same path a few moments later in pursuit of them, he halted suddenly at sight of Asa and the others returning. Charley looked crestfallen, but Billy was giggling as usual. He had not understood what they were doing, but willingly followed, supposing some game had been proposed.

"De boats all gone," said Asa, sorrowfully. "Mr. Thatcher an' Mr. Lofton must 'a' took 'em ter go ter dey traps."

"Let 's hurry back, then," said Joe, after a few moments' blank pause, "so that they won't know we tried to escape."

The run to the boat landing and back, a distance of little more than two hundred yards, had scarcely consumed five minutes, and the four spectators were again on the scene of the fight before the combatants had noticed their absence. They returned just in time to see Sweet strike the ground heavily beneath the weight of his antagonist, who now partly rose, placing his knees upon the breast of the vanquished.

"You got enough ?" shouted Bubber. "If you ain't, I kin break ever' bone in your body 'fore I quit."

Sweet said nothing, but ceased to struggle. Presently Bubber let go his hold, and rose.

"I 'll git even with you yit," said Sweet, with a black look, as he painfully gathered himself up. "You can't git away with me that easy."

The victor disdained a retort, and walked back to the camp, followed by the two boys, leaving Sweet to vent his uncomfortable feelings in threatening curses.

The round of camp life was taken up again as if nothing had happened. A week passed, during which no further opportunity to escape presented itself. Each day witnessed a gradual weakening of Joe's resolve not to make the promise required by the chief of his captors.

Thoughts of his father, his mother, his sister, haunted the boy; what would they think when they returned home and found that he and Charley had gone, no one could tell where? Had the people at home not grief and anxiety enough already? Ten days had now passed since they had gone down the river, and probably they were at home by this time. Perhaps they were even now searching for the lost boys. It was difficult to hold out, tormented by these thoughts.

The boys had been just one week on Deserters' Island, when one morning Joe said to Bubber, —

"If you'll let us go, Mr. Hardy, I'll promise you I won't guide anybody back here, or tell where you are."

"I reckoned you'd say that bimeby," answered Bubber.

"If you'll take us across the prairie and put us on the trail to Trader's Hill, we'll leave Asa and won't inform against you. It's wrong to do it," Joe added; "but I must do it on account of my mother and father; they have trouble enough without this."

Hardy was vastly amused at Joe's air of condescension, and smiled grimly. "If I was a mind to, I might devil[1] you a little," he said, "but I won't. I'll go talk it over with the boys," he added.

He did talk it over with "the boys," as he called the

[1] Cracker for tease.

other deserters; and later in the day he informed Joe that
it could not be done.   The other men refused to consent.

"I reckon you boys will have to put up with our company a while longer," Bubber said to them with a twinkle
of the eye.   "You must n't think I was jes' devillin' you,"
he added seriously.   "I 'm willin' to take your word and
let you go, specially as ther 's mighty little likelihood of
yer ever bein' able to find yer way in h-yer again.   But
the rest of 'em won't risk it."

At first Joe and Charley were very angry, the former
not hesitating to show it; but they soon cooled down, and
became very much depressed.

"Never mind," said Joe to Charley and Asa later;
"maybe we 'll make our escape before long anyhow, and
then we 'll be free to tell the soldiers where to find them."

Another week passed, — a wearying waste of time, during
which the young prisoners were a prey to growing anxiety.
They were never allowed to go out of sight of camp, except now and then to follow a deer-hunt, in the company
of half-a-dozen men.

They were not ill-treated : they were well fed ; they slept
warm and dry at night; they found some amusement in
hunting, in Billy's follies, in listening to Asa's tales and
to the deserters' yarns.   But every hour they chafed, and
were constantly proposing plans and watching for oppor-
tunities to escape.

CHAPTER V.

ONE morning about two o'clock a large animal came close to the camp, probably attracted by the refuse of a deer's carcass; and all hands were roused by the furious baying of the dogs. Snatching up their guns, the deserters to the last man sallied out and followed in pursuit. Billy ran after them, and Joe and Charley were left alone with Asa.

The eager hunters were hardly two hundred yards away before Joe and Asa looked at each other significantly across the camp-fire, now stirred to a bright blaze. They began their preparations without a word and without a moment's delay. Joe took his gun, Charley his hatchet, and Asa collected some eatables in a bucket and picked up an axe.

They were still at the fire when the sound of footsteps startled them, and a voice shouted, —

"Bubber says you all come, too. Come on, quick! Ever' las' one of ye."

The two men who had hurriedly returned on this errand halted as soon as they were within call, and waited impatiently to be joined by the negro and the boys, evidently afraid they might miss seeing the game run to earth.

Nothing but the fear that the boys might run away and betray them could have induced them to return.

The two boys and the negro exchanged glances; clearly there was no help for it. Armed as they were, they moved forward at the bidding of the two deserters, Asa delaying only to drop the bucket of food out of sight in the bushes.

The cause of the excitement, which proved to be a bear, had beaten a hasty retreat toward the centre of the island, and there, being hard pressed, climbed a tall pine. By the time the hunters reached the spot, the bear had comfortably ensconced himself among the clustering boughs at the top. Nothing could be done now until daylight, and the hunters proceeded to make themselves comfortable. Several fires were built, forming a circle around the tree, in order to make sure that the bear would remain where he was in case the watchers should fall asleep.

Then Asa was sent back to camp, in the company of two men, to bring a jug of corn-beer and something to eat. The besiegers had a merry time of it during the three hours of waiting. Even Joe and Charley forgot their disappointment in their absorbed interest in what was taking place. The treeing of a bear in a tall pine after this fashion was considered a very remarkable occurrence by even those deserters who were old hunters. Several declared that they had never seen anything like it.

" The old Okefenokee is the place to run up on curious things," said Bubber Hardy, musingly. (He pronounced the word " Oke-fe-noke.") He was lounging on the grass near one of the fires, the two boys and several of the men in his company.

"I've seen a heap o' strange things in this place," he continued, "when I use to come in h-yer huntin' before the war broke out. I reckon you boys would n't believe me, would you, if I was to tell you I seen a catfish whip a moccasin in h-yer one time?"

The men laughed incredulously, but demanded the particulars.

Bubber showed no haste to satisfy their curiosity, quietly drinking a long draught of corn-beer from a gourd passed to him by Asa. "Gim-me a chaw o' tobaccer," he then requested of his nearest neighbor, who was known as Zack Lofton.

"I ain't got none with me," was the apologetic response, which evidently failed to carry conviction.

"You never do have none *with* you, looks like to me," said Bubber, smiling rather coldly. "Lofton is about as stingy as they make 'em," he added, addressing the others. "I believe he'd skin a flea for its hide and tallow."

Lofton did not enjoy the general laugh which greeted this pleasantry; and if any other man present but the "cock of the walk" had uttered it, he would have given him the lie very promptly. As it was, he contented himself with retorting in an injured tone, —

"You've chawed a heap o' my baccer, Bubber Hardy."

Being provided by some one else with the desired "chaw," Bubber proceeded to tell his story. It was, in substance, that he had once seen a moccasin spring upon a catfish in a shallow lagoon of the swamp, and promptly get "whipped;" that is to say, disastrous consequences resulted from the snake's attempt to swallow its prey.

For the fish immediately "popped" its formidable fins through the reptile's throat, and all efforts on the part of the latter to disgorge its victim proved futile.

"I depend that moccasin reared from away back and was as vigeous a snake as you ever laid eyes on," Bubber declared, with a laugh; "but it bit off more 'n it could chaw, shore enough, that time."

He wound up by saying that the snake crawled off rapidly out of sight; but several hours later, returning past the same neighborhood, he found it lying dead, the tail of the fish still protruding from its mouth and the fins visibly transfixing its neck. The catfish still lived, and Bubber was induced to go to the trouble of liberating it.

Hardy's listeners had expected a jest, but they accepted the story as matter of fact, or at any rate as probable enough; and no one presumed to give expression to doubts, if any were felt.

This was the beginning of much spinning of yarns, — some of them quite remarkable, — which amusement, with intervals devoted to jesting and discussion on suggested topics, was kept up until daylight.

"That ain't ez strange as some things I've seen in the Oke-fe-noke," said a thin wiry little man known as Bud Jones. Although a white man, and apparently not lacking in "hard" common-sense, he was noted for his firm belief in witchcraft.

With this introduction to his tale, he went on very seriously to relate how a charmed deer had come "right up" to him three times in the swamp one day, and how he had tried to shoot it down in vain. He assured his hearers that if he could have moulded a silver bullet and shot that, he would have been successful. He had heard that the like had been done in a similar case, but admitted that the authorities differed as to the result; some said that the charmed deer had thus been brought to earth, but others claimed that upon the discharge of the mysterious silver bullet the animal had vanished away "right there in the broad open day."

In proof of the reality and efficacy of charms, Bud Jones related further how once, long ago, his own mother had sickened, and was afflicted with great fear of a certain old woman in her neighborhood; how, at last, some one advised her to wear red pepper in her shoes, and, having done so, how she promptly recovered her health and was relieved of all further apprehension.

Bubber Hardy and most of the others smiled incredulously at this story; but Asa listened with a solemn face and absorbed attention.

"Dat's de trufe, you year me," he declared in a low earnest aside to Joe. "A 'oman put bad mouf on me dat-a way one time, and I tell you she everlas'nly gim-me de devil, too."

"You ought to have more sense, Asa," was the unsympathizing reply.

"Well, Joe, what's the strangest thing you've seen in the Oke-fe-noke?" asked Bubber, after several other men had related more or less startling experiences.

The boy felt like replying, in substance, that the strangest, most unaccountable, most infamous sight he had seen in the great swamp was a party of able-bodied men in hiding, once called soldiers, who had deserted their posts of duty in the hour of greatest need. Prudence, however, restrained him.

"I have n't seen as much of it as the rest of you," he said modestly, after a moment's thought; "but the strangest story about it I ever heard was the one father said the Indians used to tell a hundred years ago."

"Less hear it," cried several.

So Joe, after his own boyish fashion, related the old Indian legend which pictured the remote interior of the Okefenokee as a high and dry land, and one of the most blissful spots of earth, where dwelt beautiful women called daughters of the Sun. Some warriors of the Creek nation, lost in the interminable bogs and jungles, and confronted with starvation and despair, were once on a time rescued and lovingly cared for by these radiant creatures. And ere the belated warriors were led out of the confusing labyrinths and sent on their way, they were fed bountifully with dates, oranges, and corn-cake. There may have been other good things, but Joe's memory could vouch for only the dates, oranges, and corn-cake.

Joe remembered that his father had said it was a pity that ambrosia was not substituted for the last item. Corn-cake is doubtless a good and useful thing in its own way ; but something a little less commonplace would seem more fitting in the realm of legend. The maize, however, was probably regarded by the Creek Indian as one of the most precious and useful gifts of the gods, and therefore not unworthy of a place in this legend of the daughters of the Sun who dwelt in the great Okefenokee.

The deserters one and all seemed interested in the story, and paid Joe the compliment of inviting him to tell another, — an invitation which he modestly declined.

The fires were now replenished, further draughts of beer were drunk, fresh pipes were lighted, and the spinners of yarns began another series devoted to the "tight scrapes" in which they had found themselves occasionally in the Okefenokee. One man told of a deadly hand-to-hand conflict with a wounded bear; another of a thrilling unarmed fight with a wild-cat; a third related how he had once sunk down suddenly to his armpits in the "prairie," how he had saved himself by grasping the growth on a small tussock within arm's length, and how he was confronted there, before he could drag himself out, by an angry moccasin, which luckily he shot.

"Talkin' 'bout tight scrapes puts me in mind o' the time I went tiger-huntin' with Seth Mixon," spoke up Zack Lofton, who had said little till now.

"*Tiger*-huntin'?" repeated Joe, in astonishment. "There are no tigers in this country."

"I depend if I ain't seen a tiger in this swamp more 'n once in my time, and ez survigeous a beast ez you want to run up on, my name ain't Lofton!" was the emphatic response.

"Some calls 'em tigers, but they ain't nothin' but panthers," explained Bubber Hardy.

"Me and Seth come in at the Pocket on t' other side," continued Lofton, referring to a peninsula extending about ten miles into the Okefenokee on the western side, near the point where the famed Suwanee emerges from the

4

great swamp, its mother, a sluggish little river of dark, wine-colored water. "It's fifteen years ago now and better. We got in ez fur ez Billy's Islant by night and camped there; and that night we heard a curious hollerin' in the swamp that sounded a little like a poor-job, and a little bit like a cryin' baby, and we knowed it must be a tiger."

Accordingly, very soon after breaking camp next morning they saw panther signs. But the dog soon lost the scent, the panther, like the wild-cat, being accustomed to traverse the jungle less perhaps afoot than on high among the interlacing branches. It was now proposed that the two hunters separate, agreeing to hail each other after a certain length of time.

Half an hour later, as Lofton stole guardedly through a clump of tall bushes and, thus screened, looked across a small, comparatively open space, he observed a curious agitation of the underbrush about fifty yards distant. Leaping to the conclusion that the panther was there, and deciding to risk a chance shot, he raised his gun, and was taking aim at the swaying branches, when he received a sudden stunning blow which knocked him off his feet. As he fell, he was conscious of the sound of a gun's report, and of a stinging, tearing pain in his right shoulder. He had been shot.

"I groaned and kicked powerful lively," said Lofton, with a grim smile; "but when another load o' buck-shot came a whistlin' thoo them bushes right over my head, I laid there mighty still. I gethered my gun, though — with my left hand — and ef I could 'a' got at that blasted fool,

Seth Mixon, right then, he'd 'a' h-yeard from me. I was so mad I a'most believed he shot me a-purpose."

He was, however, prudent enough to call out; and the horrified Mixon ran to him, protesting that he thought he was shooting the "tiger"!

"I come mighty nigh makin' the same mistake, but I got fightin' mad all the same," Lofton declared, with something of regret; "and I depend I give Seth Mixon a piece o' my mind that day. I got up and tried to walk home, but had to lay down agin', and kep' gittin' weaker and weaker. Mixon said the only thing to do was for him to go and git a horse and put me on it, but I'd have to lay there the best part of a day 'fore he could git back. I told him to cut out, and off he went, blazin' the trees behind him. It want long 'fore I felt sort o' numb like, and dreckly I sort o' dozed off — fainted, I reckon. I'd clean forgot about the tiger; but when I come to I 'membered it agin mighty quick.

"I knowed sump'n was up soon ez I seen my dog fidgetin' aroun' and whinin'. The hair was up straight on his back, and his tail was 'tween his legs. But soon as ever he seen me stirrin' he showed more spunk, and commenced to bark at some thick brush 'bout forty foot off. I knowed right off it was the tiger, and that it smelt my blood and was after me. Dreckly I seen its tail workin' back and forth up on a high limb, and I knowed it was fixin' to jump."

Observing that every one around the fire was listening intently, Lofton took a fresh chew of tobacco, and went on to tell how, for some little time longer, he lay perfectly quiet, fearing that the slightest movement would be the

signal for the attack. At length, unable longer to bear
the suspense, he partly raised himself up, grasping his gun
with his left hand.

The moment he did so, the panther tore through the
obstructing branches with a horrible growl, and sprang at
him. But the distance was too great, and the beast struck
the earth some ten feet away. Before it gathered itself for
another leap, the dog had sprung forward in defence of his
fallen master.

"Then they had it, nip and tuck, tooth and nail, and sich
howlin' and snarlin' I never h-yeard in all my born days."

Struggling to his knees, Lofton managed to cock his gun
and raise it to his shoulder with his left hand and arm; but
he hesitated to pull the trigger, fearing to shoot his dog.
For the two animals, fighting to the death, were never still
for one moment, now here, now there, backward and forward, now rolling over on the ground and gradually nearing
the wounded man. In a short time it was evident that the
dog was failing.

"I was shore my time hed come," said Lofton, solemnly,
"but I held my gun and watched my chance. They kep'
a-comin' closer, a-wheelin' roun', and dreckly the tiger made
a jump and fetched herself and the dog with her in three
feet o' me; and, sir, I leaned over quick ez a flash and put
the muzzle o' my gun right spang aginse the back side o'
her head, and blazed away.

"Well, mebby you won't believe it, men, but that cat
jumped right straight up ten or twelve foot high, jerkin'
loose from the dog. When I seen her comin' down, look to
me like she was comin' right for me, and I sort o' give up

and went off agin. You see, I'd been bleedin' like a hog, and was mighty weak.

"And what you reckon ? When I come to, the tiger was layin' dead on one side o' me, and the dog putty nigh dead on t'other. I thought he *was* dead at first, and I a'most broke down and cried. I dunner how long I laid there; I didn't have no sense left scacely, and it was a mighty good thing another tiger didn't come along that day.

"By and by I heard 'em comin' thoo the swamp; and when they got to me and lifted me up and put me on the horse, and one o' Seth Mixon's boys says, 'Shoot that dog and put him out'n his mizry,' I up an' spoke a piece o' my mind, and I made 'em strap that dog on the horse behind me 'fore I was done.

"Well," Lofton concluded, gazing absently into the fire, "they got us home, and we both got well atter a while, — me and the dog; but ther's a buck-shot or two in my shoulder yit, and sometimes it hurts me so I kin scacely strike a lick o' work. You mer say what you please, but that was the tightest scrape I was ever in."

"So that's what makes you lame in the right arm ? I always thought you got that wound in the war," remarked Bubber Hardy.

Joe and Charley were both intensely interested in this story, accepting it without question; but the former now noted a slightly sceptical expression on the face of the "cock of the walk," who evidently cherished no admiration for Lofton.

"Day's a-breakin'!" some one called out at this moment;

and the loungers about the fire sprang to their feet, turning their eyes toward the top of the pine, where the bear had taken refuge.

As soon as there was sufficient light to outline the black bulky form among the branches, the hunters opened fire, one at a time, and at the thirteenth shot the big game came tumbling down, striking the earth with tremendous force.

The bear measured seven inches across the ball of the foot, and three inches through the fat on the round, and the total weight was calculated at not less than four hundred pounds. The skin was carefully taken off, many pounds of the choicest meat sliced to dry, and the rest of the carcass left where it was for the vultures. When the sun was some two hours high, all hands, in great good humor, returned to camp and partook of the hot breakfast which Asa had now prepared.

CHAPTER VI.

AFTER eating a heavy breakfast, most of the deserters
lay down on the grass in the shade and went to sleep.
Joe, too, felt drowsy after the unwonted loss of sleep occa-
sioned by the bear-hunt, and presently followed the exam-
ple of his captors. Thus Charley and the half-witted Billy
were left alone with Asa, who busied himself washing the
pots and pans over the fire.

"We had *such* a good chance last night," remarked
Charley, regretfully, — "if only they had n't remembered
and sent for us. Did n't we, Asa ?"

"Num-mind," said Asa, consolingly; "we 'll git another
chance. Some dese days dey 'll clean fergit us, an' we 'll
gie 'em de slip. We 'll lead 'em a race some dese days."

"A chance to run a race ?" asked Billy, vaguely. "Is
that what you want ? I 'll run a race with you right now."
His vacant eyes quickened with a sudden enthusiasm.

"We did n't want to run a *race*," answered Charley, dis-
couragingly.

Suddenly the half-witted young man leaned over toward
Charley, and said to him in a low voice, with the air of
one conferring a priceless favor, —

"Would you like to come now and see son ? Say, boy ?"

"Who is 'son'?" asked Charley, curiously.   "Yes; I'd
like to see him."

"Come on, then."

Asa was now engaged in vigorously scraping one of his
pans, and did not overhear this.   When he looked up again
from his work, Charley and Billy had risen and walked
away.   The latter, who had fished out of his pocket a
small wriggling frog and carried it in his hand, led the way
through the woods about a quarter of a mile, halting at
last near the clay-covered roots of a large pine that had
fallen during a storm.   At the base of this was a small
round hole in the earth, and here Billy fell on his knees,
and began repeating in a strange, monotonous, coaxing
voice, —

"Doodle, doodle, come out your hole !   Doodle, doodle,
come out your hole !"

These are the mystic words popularly believed among
the children of the Southern States to be potent to call
forth from ambush the ant-lion, which crafty insect pre-
pares over its nest a kind of pitfall for ants.   Charley saw
at a glance that this was no ant-lion's pitfall.

"That's not a doodle-hole; that's a snake's hole," he
exclaimed, stepping backward.   And indeed the hole was
hardly less than two inches in diameter.

Billy made no reply, and continued without intermis-
sion his peculiar recitation of the supposed charm.

"I hear him a-comin'," he said softly, at last.   Then, in
a gentle, caressing voice, he continued, "Come on, son ;
come on, son."

In a few moments a large rattlesnake glided out of the hole, and seized the frog from Billy's fingers. Charley backed rapidly away, and sprang upon a log, but Billy did not move from his place, and showed no fear whatever.

"Come away from there!" cried Charley, all amazement. "You Billy — that snake will bite you!"

"Son won't bite me," replied Billy, confidently. "Son knows me. You neenter be a-scared, boy; son won't hurt you if I tell him not to."

So this was "son," — this was the great mystery which poor Billy had seemed so to delight in!

"If you don't come away, I won't stay here," cried Charley, urgently. He was really alarmed for Billy's safety, being convinced that as soon as the snake had swallowed the frog, the foolish boy would be bitten.

After begging him again and again to come away, Charley jumped down from the log and hurried back to camp. He thought he ought to inform Sweet or Bubber at once, but they were asleep; and by the time he had detailed the story to Asa, the witless snake-charmer himself appeared unhurt.

"Lem-me tell you one thing," commented Asa, with a serious face, as soon as Charley had made him acquainted with the facts: "you let dat Billy hoe his own row. Play wid 'im roun' de camp much ez you like, but don't you go foolin' long atter him roun' dese woods. He ain't got good sense, an' he 'll git you inter trouble sho 's you born."

"Look yuh, Billy," he asked, as the latter approached and took his place at Charley's side, "ain't you got no better sense 'n ter prodjick wid a rattlesnake dat-a way?"

"What made you tell?" asked Billy, reproachfully, of Charley.

"Dat snake goin' to bite you an' kill you ef you don't mind," continued Asa, severely.

"Don't you fret," said Billy, giggling immoderately. "Son knows me."

"When they were all tellin' stories round the fire this mornin', why did n't you tell one, too, Asa?" asked Charley, when the subject of Billy's snake had been dropped.

"Nobody did n't ax me," replied Asa, with a guffaw. "I could 'a' tole 'em 'bout how a 'oman put bad mouf on me an' kunjud me one time, but dey did n't ax me."

"Well, you can tell us now, can't you?"

The negro was by this time beginning his preparations for dinner. He now sat on the grass near Charley, rapidly removing the feathers from a wild turkey which one of the men had shot on the previous day.

"I ain't got time nohow," he replied; "but ef I was to tell you any tale I ought to tell you dat 'n Unker Tony use ter tell de chillun 'bout de tuckey-gobbler an' de rattle-snake. De way Billy been foolin' wid dat snake dis mornin' put me in mind o' Unker Tony's tale, an' ef he don't look out he gwine to come out like de tuckey did, too."

"Oh, tell it; you 've got time," urged Charley. "May-be it 'll make Billy have more sense."

"'T ain't no great tale," said Asa, by way of introduction, having been induced to begin. "Hit 's des a tale to tell bigity chillun when dey git too mannish. Unker Tony say one time, way back yonder, when de tuckey use ter be de mose proudes' bird in de woods, a ole tuckey-gobbler 'uz

comin' long, an' fuss ting he know he run up on a rattlesnake. De tuckey strut long so bigity, wid he tail spread
out so fine an' he head reared up so high, he did n't hardly
see de rattlesnake, an' look like he gwine walk straight
on over him.

"Rattlesnake shake he rattle — *z-z-z-z-z-z !* — an' he say
'Don't yer walk on me; don't yer walk on me dis
mornin' !'

"Tuckey-gobbler look down at 'im out de cawner he eye,
an' he say : 'Eh? Was you speakin' to me ?' Den he
look hard at de rattlesnake an' mek out like he so little he
don't know 'im, an' den he turn up he nose an' laugh to
hisself an' come a-walkin' right on.

"Rattlesnake bristle up an' ,squirm roun'; he say,
'Don't you walk on me. Bet'ner walk on me; I tell you
in time — *z-z-z-z-z-z-z !*'

"Tuckey-gobbler say, 'Humph! ef sich a triflin' lil
wurrum like you so partic'lar, I tink you better git out de
road.'

"Rattlesnake shake he tail wuss. He say, 'You mus'
be crazy, enty ? I 'll have you to understan' I don't git
out de road fer nobody, let 'lone fer sich a stuck-up fool ez
you is !'

"Ole tuckey-gobbler rear back an' say, 'Who is you, I
like to know, to be talkin' yuh so bigity ? You little 'sig-
nificant bug! Is you got de enshoance to stan' dere an'
sass *me ?* You don't know me, does you ? You dunner no
better 'n to lay under dat bush an' shake yo' tail at ME ? —
when — my — gran'daddy — swallowed — a ALLERGATER !'

"De ole tuckey stretch hisself up powerful big an' look

like he believe *he* could mose swallow a elephant. An' de rattlesnake bust out in a big laugh, an' he up 'n say, —

" ' Dass you, is it ? I said to merself you was a fool when I fust seen you comin'.' An' den he laugh fitten to bust.

" Ole tuckey-gobbler git fightin' mad, you see him so, an' he say, 'Shut up dat! I 'll make you laugh on t' other side yo' mouf turreckly. I aim to make you eat dem words 'fo' I quit,' an' den he wheeled in an' everlas'nly cust de rattlesnake out.

" Rattlesnake shake he tail fast ez lightnin' — *z-z-z-z-z-z-z !* He say: 'I dare you to walk on me! I des dare yer — double dog dare yer — to walk on me ! '

" An', would you b'lieve it, de ole tuckey so mad he des up 'n pounced right on de rattlesnake an' tried to pop he spurs in him; but de rattlesnake done bit him — *dat quick !* [Asa snapped his fingers loudly.] An' little more, an' dat bigity tuckey-gobbler done drap down dead."

" Now you see that, Billy," exclaimed Charley, and Asa shook his head in solemn warning.

But Billy did not appear to be in the least disturbed, responding with his usual giggle.

" Oh, but you see, that was n't son," he said argumentatively ; " that was son's cousin, I reckon. Son won't bite me. No-sir-ree ! "

CHAPTER VII.

AFTER supper that evening, as the men told yarns and joked about the camp-fire, Billy seemed unusually wide awake, and repeated nursery rhymes and rigmaroles by the dozen.

Taking Charley's hand in his, he touched the fingers one after another, repeating, "Little man — ring man — long man — lick pot — thumpkin."

Then, tweaking the toes of his own bare feet, he merrily recited, —

> "This little pig wants some corn ;
> This one says, ' Where you goin' to git it ?'
> This one says, ' In master's barn ; '
> This one says he 's goin' to tell ;
> This one says, ' Queak ! — queak ! —
>   Can't git over the door-sill ! ' "

Touching first Charley's index finger and then his own as each word was uttered, he said, "William Ma-trimble-toe; he 's a good fisherman; catches hens, puts 'em in pens; some lays eggs, some lays none; wire, brier, limber lock; sets and sits till twelve o'clock; O-U-T spells ' out ' — go !"

This suggested a game of hide-and-seek, and Charley was coaxed into playing. Before long Asa joined, and then Joe was drawn into the game. It was a bright moon-lit night, and no one seemed sleepy. The deserters stopped telling their yarns, and watched the game. The laughter of the boys and the negro affected them pleasantly.

The fun was contagious. Five minutes later every oc-cupant of the island was engaged in the sport. One by one the deserters yielded to the fascination of it, and joined in the game, surprised at themselves and at each other, but excusing such levity with the laughing remark, "Anything for a little fun!"

"Ten — ten — double ten — forty-five — fifteen hun-dred — are you all hid? Are all my sheep hid?" shouted Billy; and such whooping and running and hiding in far dark recesses as followed!

"Now's de time!" whispered Asa, when the fun was at its height, and he and Joe and Charley had run off and squatted together behind the same clump of bushes. "Now's de time for us to give 'em de slip an' git away."

The boys listened eagerly as he explained the plan which he had formed during the past few minutes. He proposed that as soon as the players scattered to hide the next time, he should run off to the boat-landing, step into the water, and drag each of the three boats about a hun-dred yards off into the submerged forest, where they could not be found readily. The deserters would thus be led to believe that the boys and the negro had escaped to the prairie, taking all the boats with them.

While he was hiding the boats, the boys should continue the game, showing themselves conspicuously, in order that the absence of Asa, if observed, might not excite suspicion.

The negro had outlined his plan thus far, when the course of the game compelled the conspirators to separate and return to headquarters. When the rush for cover was again made, the boys saw Asa dart away in the direction of the boats, and were well pleased to observe that his absence attracted no attention. They were then careful to keep suspicion lulled by playing with all their might. The cunning negro succeeded in secreting the boats as proposed, and in a very short time turned up again, none but the boys observing that his ragged trousers were wet to the knees.

Joe and Charley understood that the first rush for cover after Asa's return was the time to escape. When they saw him again dart away along the path into the swamp-cane, they followed fast with throbbing hearts, arriving at the boat-landing by the time the last one of the scattering men was safely hidden.

There Charley was given his hatchet, and Joe his gun. Asa put a rifle over his own shoulder and snatched up a bucket of eatables, — all of which he had cleverly conveyed thither since the commencement of the game. Asa stepped into the water, and bade the boys follow.

"We got to go in dis water to fool dem dogs," he whispered.

He led the boys about fifty feet from the shore along the open boat-road, then turned to the right into the thick growth, and skirted the island for several hundred yards before landing again. It was no trifling undertaking. The water was in many places over their knees, and was thick with drift and moss; the bottom was often muddy, and a dense swamp undergrowth forced them to a tortuous route. Besides, little light descended from the moon among those crowding trees. Poor Charley found it difficult to keep up.

"Ten — ten — double ten ! " they heard Billy shouting faintly as they landed, and knew that as yet no one observed their absence.

Asa had not dared to risk flight across the prairie without Billy to carry the third and last boat. Even two boats would have been more than they could move in rapidly enough to escape pursuit and capture.

He had, therefore, decided to secrete the boats, putting the deserters on the wrong scent and causing delay. After covering their trail in the water, Asa meant to strike across the island and enter the swamp at the opposite end. He knew there was a way out of the Okefenokee through a jungle in that direction, which could be followed on foot, though he had never been over it.

"Whose rifle is that, Asa ? " asked Joe, as they started forward in single file.

"Bubber's," was the answer, with a low laugh; "I aim to take dis rifle to yo' pa ter pay 'im fer my rent, — fer de five months I been workin' fer Bubber."

"He won't have it," replied Joe, "and you ought not to have taken it."

"Ef he don' want it, den hit 's mine."

Joe laughed in spite of himself, and they all halted to listen as a shout reached them from the camp. Distinctly they heard the names of Joe and Asa called, and knew that they were missed. They now went forward faster than before. Five minutes later another shout reached them; and after a brief silence several sharp short yelps from the dogs were heard.

"They have found that the boats are gone, and have called out the dogs," said Joe.

Asa leaped forward at the sound, and poor Charley was hard pressed to keep up after that. The darkness was bewildering until they emerged from the "hammock" and started along the open pine ridge which was the backbone of the island. Here, however, the moonlight filtered through the scattering tops of the pines, and they could distinguish prominent objects fifty feet away.

Nevertheless, it was very difficult to make rapid headway, owing to the frequent blackjack thickets, the tall huckleberry and gallberry bushes and the crowding clumps of fan-palmetto which barred the way. There was a slight trail leading down the ridge, as they knew; but no time could be lost in searching for it now, and they were obliged to pick their way as best they could.

The island was about four miles long, and fully an hour and a half was consumed in covering the distance. Descending into the dense growth of the "hammock" which joined it with the swamp at the farther end, they halted to listen. All was deathly still, at least in the direction of the deserters' camp; but the stillness of the dark, slumbering swamp in their front was suddenly broken by the dismal and unearthly hoot of an owl.

Joe thought they ought to push forward and make good their escape into the swamp before daybreak; but Asa's courage now failed him, and he objected. He said it was dangerous to go on, as indeed it was; they might sink into the bog over their heads, or they might be set upon by a panther. Besides, there was no telling what sort of reptile they might stumble upon in the darkness. Joe was by no means free from fear himself; but he thought it the part of prudence as well as of manliness to advise going forward.

" Dem men won't start to hunt us 'fo' daylight," said Asa, confidently. " It's midnight now. Dey t'ink we gone on de prairie wid all de boats, an' I know mighty well dey ain't gwine start wadin' atter us till mawnin'."

They stood a moment in silence. Suddenly from the dark depths of the swamp on their front a strange cry was borne to them, — a cry or bark or croak, they could not tell what.

" That sounds like a bear," whispered Joe.

" Must be a pant'er," whispered Asa.

The cry was heard again, more mysterious, weird, and startling than before. Facing about, they retreated hurriedly up the slope and into the open pine woods, where the moonlight outlined neighboring objects.

Asa, badly frightened, wanted to build a fire, but Joe would not consent to such an imprudence, and finally it was agreed that they sit down with their backs to a large pine and watch until daylight.

Joe and Asa sat thus, upright and alert, their guns in readiness, for a long while. Charley lay down between them, and fell asleep. All was now quiet, and gradually they recovered from their fright.

Gradually also a drowsiness seized them. Asa rested

his gun across his lap, dropped his head on his breast, and soon began to snore. Joe roused him several times, only to see him lapse into insensibility a few moments later. The boy watched more than an hour longer, and then he also succumbed. Later, as he roused up to a state of semi-consciousness and opened his eyes, he saw that the moon was low, and that apparently all was well. However, as he drifted back toward dreamland he thought he heard a short, sharp yelp or two from dogs in the distance, but was too much enchained by drowsiness to heed.

The dogs had started some trail, no doubt, — that of a rabbit, perhaps ; but what could it matter to the three sleepers under the pine ?

When Joe again awoke it was daylight, and the dogs were leaping about him and barking. Several men were at hand, too ; and the one nearest, who looked down at the sleepers with an ugly grin, was Sweet Jackson. The sound of blows then drew his attention to the fact that Bubber Hardy, close by, was kicking Asa in order to awake him.

They were caught ! What else could they have ex-pected ? The events of the night leaped into view in the boy's memory, and he was overcome with sorrow and shame. " If we had not been such cowards ! " he thought.

Joe rose to his feet. Charley was crying, and Asa was looking around stupidly. Sweet laughed in derision as he looked at them, and poor Joe thought that even this was deserved. After some severe kicking and cuffing, from which Joe and Charley turned away their indignant eyes, Asa was allowed to rise.

"You thought you 'd run off with them boys, and steal
my gun in the bargain, did you?" shouted Bubber, angrily.
"I 'll make you sick of it 'fore I quit."

The boys themselves were seized roughly, and all were
marched back to camp. Asa was ordered to cook breakfast,
and the men immediately set about building a prison, — a
sort of pen of heavy saplings, with slanting poles and a palmetto thatch for the roof. It was given no window and
only a small aperture for a door. At night Asa was shut
up in this pen, but Joe and Charley slept in the loft with
Bubber as formerly.

The boys found themselves under constant watch after
that, and their freedom of coming and going in the neighborhood of the camp was curtailed. Still it was not impossible for them to get a private word with Asa while he
was doing his work; and one day some three weeks later
Joe said to him, —

"If we could catch a live duck, maybe we could send a
note to father."

"Dat duck would n't go to yo' pa," replied Asa, stolidly.

"Well, I heard Sister Marian and Captain Marshall
talking about a book they had read," continued Joe, "and
they said a lady and a man she did n't like were cast
away on an uninhabited island in the Pacific Ocean, and
after a while the lady began to love the man because he
was so kind, and did so many things for her. And one of
the things he did was to go out before daylight and wade
into a pond where ducks came very early in the morning;
and he would squat down in the grass and water up to his
neck (the grass hid his head), and when the ducks swam

close up, he would reach under and grab them by the foot. And he would write letters, telling how the lady was ship-wrecked on that island, and tie them to the ducks and turn them loose."

"An' did dem ducks carry de letters to the right place?" asked Asa.

"I don't know. But after a while a ship came. I'd like to try it, anyhow."

"We can't git no ducks," said Asa; "but if you write de letter, I'll git you a live pa'tridge."

The deserters had set traps for partridges at several different points on the island, and had usually a supply of birds alive in a pen near the camp, with which to vary their diet. The flight of a partridge seemed to promise less than that of almost any bird they could think of; but it was the only chance, and Joe accepted the suggestion.

So when Asa went to the bird-pen the same day and wrung the necks of a dozen partridges, he brought back with him a live one also, and turned it over to Joe without attracting attention.

Joe, having written the letter, tied it securely beneath the bird's wing. It ran : —

DEAR FATHER, — Charley and I got lost in the Okefenokee, and we came to this island where the deserters stay. They keep us prisoners, for they are afraid we will tell the soldiers where they are.

Asa is here too. They stole him. We tried to escape two or three times, but it's no use.

When you come after us don't forget this, — that to the north of this island there is a great wide marsh, I don't know how many miles across, and beyond it, they say, is a trail that goes to Trader's Hill.   Come quick.

Your affectionate son,

JOSEPH MÉRIMÉE.

P. S. — Whoever finds this, please take it or send it right away to Mr. Roger Mérimée, Trader's Hill, Charlton County, Georgia.

The letter was written on a page torn from the boy's notebook.   Fastening it beneath the bird's wing, and tying about its neck a strip from his handkerchief to attract attention, Joe pitched the partridge upward with all his might, hoping thus to frighten it into a long flight across the prairie.   He knew that if it alighted on the island the chances of its being shot or caught by a friend would be altogether lost.

The bird soared high, plunged, wheeled at two hundred yards' distance, rose again as if newly alarmed, then quickly dropped into the island jungle.   Joe sat down and buried his head in his hands.   For the moment all his hopes were over.

"What you let that bird go for, you triflin' nigger?" cried one of the deserters on watch.

"Dass a mighty smart bird.   He ain't want to lose his fedders," said Asa, grinning; for he knew the loss of one partridge was nothing to the deserters.

"Don't you cry, Mas' Joe," he whispered, bending over the boy; "you done yer bes'.   Mebby we find some other way."

But Joe's hopes had been so high that he could not soon control his silent tears.

CHAPTER VIII.

JOE AND CHARLEY COVER THEMSELVES WITH GLORY.

BY this time the deserters had begun to relax their vigilance; and the two boys were allowed to walk about the neighborhood of the camp with almost their former freedom.

"They can't git away without Asa's help nohow," Bubber Hardy more than once remarked to his associates; "and as long as we keep our eye on the nigger we're all right. No use hemmin' the boys in too close. It's hard on them, powerful hard; and I, for one, don't like to see 'em suffer. You kin see all the time they're bad off with homesickness; and they air two as smart and honor'ble and well-behaved boys as I ever laid eyes on."

Such remarks, delivered now and then by the "cock of the walk," produced a perceptible effect on every member of the camp, except perhaps Sweet Jackson; and the boys soon discovered that they could go about as they pleased without molestation.

"Can't you think of some other plan for us to get away, Asa?" asked Joe, a day or two after they had let fly the partridge.

The two boys stood looking on while the negro cooked dinner. The deserters were all out of earshot.

"I reckon der ain't but one way," replied Asa, punching the fire slowly and meditatively, " an' dat's fer you boys to keep 'wake tell late some night, den slip down out de loft widout wakin' up any dem mens an' let me out dat pen. Den we kin git in one dem boats, atter we done sot fire to de yuther two, an' — "

"Set fire to the other two ? " exclaimed Joe.

"Dass hit; dass de ve'y thing to do. Den dem mens 'll sho' sweat 'fo' dey cotch us agin."

"But that would leave them prisoners on the island," objected Joe.

"And they might starve," said Charley.

"Shoo!" cried Asa, with a laugh of absolute indifference. "Dey mout ez well be prisoners ez for us to be prisoners. Des be turn an' turn about. You said yo'self dey 'zerve to be shot fer desertin' fum de army, — why can't dey starve, den, ef dey ain't got sense enough to git away from yuh widout de boats. Let 'em *root-hog-or-die*, *I* say."

"Oh, but that would be mean," said Joe, shaking his head. "I would n't be willing to do that."

"Wait tell dey kick you an' cuff you roun' lak dey done me, an' you'd be willin' to burn dey house down, let 'lone dey boats."

"No, I would n't."

"Well, den, we kin leave de boats, — des hide 'em lak we done t'other time."

"Oh, yes, that's what we can do," agreed Joe.

"An' ef we git started by midnight we be out dey reach 'fo' mornin'. Dey never kin ketch us."

It was suggested and agreed on that the attempt be made that night. The boys were warned by the negro to remain awake, and not stir from their places until they had listened for a long while to the snoring of the deserters, and were absolutely sure that they were all sound asleep. Then they should steal guardedly along the wall until opposite the door in the floor.

" Of all you boys do, don't you step on none dem mens foots dere in de dark," warned Asa, " fer ef you do de cake's all dough."

He added that it would be unwise to attempt to let the ladder down ; they had better jump lightly to the ground instead.

" Oh, we 'll just swing down by our hands and drop on our feet," said Joe. " It 's not high."

Asa said that if for any reason they failed to get off in the boats, they could run down to the other end of the island as before, and start off afoot on the jungle trail of which he had heard the deserters speak. The trouble was that he was not sure just where to find it.

Joe proposed that he and Charley spend the afternoon looking for it ; and as soon as they had eaten their dinner, and the deserters had scattered, he sauntered away, gun in hand, followed after a few minutes by Charley. If this move was observed, it excited no apprehension, and the boys got off unchallenged.

After walking about two miles down the backbone of the island, the boys concluded to cut across to the swamp on the right, and begin looking for the jungle trail. Their plan was to follow as nearly as possible the line of demarc-

ation between the swamp and the high ground, thus encir-
cling the island in the course of time, and necessarily
crossing the trail. The slight trail which led from the
camp down to the opposite end of the island was there
lost, before it entered the swamp, as they had ascertained
on a previous excursion, and it was useless to follow it a
second time. Their present plan, of following the rim of
the island, seemed the only one involving the thorough
search which they wished to make.

The path chosen was difficult to follow. Often a détour
higher up on the island, or deeper into the swamp, was
necessary to avoid bogs, marshes, impregnable clumps of
fan-palmettos, and tangled masses of brambles. And often
the way was difficult enough by reason of the aged fallen
logs thrown criss-cross, and piled high by wind storms, and
by the crowding swamp undergrowth, and the thickly
standing trees themselves. Once they penetrated a cane-
brake through which it would have been impossible to go
but for passages evidently made by wild animals; for the
tall strong reeds, which stood as straight as arrows, were
for the most part hardly three inches apart.

Even along the borders of the comparatively open pine
land which formed the island, they were most forcibly
reminded of what a wild, pathless wilderness the great
Okefenokee really was.

Two or three times they halted and carefully examined
faint suggestions of a trail, soon pushing forward again
unsatisfied. They had passed the extremity of the island,
and were returning up the left-hand side, in great fear lest
their efforts should be altogether fruitless, when they at

last came upon what Joe felt convinced was the object of their search.

Having followed the trail two or three hundred yards into the jungle, they retraced their steps to higher ground. It was now late in the afternoon and time they were turning their faces toward camp; but they had begun to feel weary after the long and rough tramp, and Charley begged that they might stop and rest. So they lay down on the soft billowy wire-grass in a high and dry spot, hemmed in by tall clumps of palmettos.

"Oh, is n't this nice!" exclaimed Charley, after a sigh of great satisfaction.

Joe was about to utter a response, when all at once they heard a rustle in the grass to the left, and the next moment the hearts of both boys began to beat with strong excitement, as their eyes fell upon a large wild-cat crouched within a short distance of them.

Involuntarily they sprang to their feet, whereupon the cat's hair stood on end, its eyes flashed with rage, and it displayed its glistening teeth, uttering a low guttural growl. The animal, which they must have surprised close to its lair, as otherwise it would likely have made off without show of fight unless attacked first, was a powerful one, some three feet in length, its hair being of a dark brownish gray, mottled with black.

Joe snatched up his gun, took hurried aim, and fired. But he was trembling with excitement; and it was no wonder that the load of buck-shot buried itself in the grass a foot or more wide of its mark.

There was no time to reload, for a moment later the

enraged wild-cat leaped through the air, landing full upon the boy ere he could spring aside. The shock carried him to his knees, the now useless gun slipping from his grasp. As the cat came down, it cruelly clawed the boy's left shoulder and the left side of his head, snarling furiously and blowing its hot breath into his face. Joe beheld its fiery eyes only a few inches from his own, and his hands flew to its throat.

Exerting all his strength, he held the cat off, but could not prevent his clothes from being torn to shreds by its strong white claws, and painful wounds being inflicted upon his arms and body.

For a few moments Charley stood paralyzed with fright, then he caught the cat by the tail and strove frantically to pull it off his brother. Failing utterly in the attempt, he thought of his pocket-knife, and getting it out as quickly as possible, stabbed the creature twice in the back. Then, with a maddened snarl, the cat let go its hold on Joe, and turned upon Joe's rescuer.

"Help! help me, Joe!" cried Charley, terrified.

"Grab him by the throat!" shouted the elder boy, staggering to his feet, half blinded by the blood which covered his face.

Joe's first thought was to seize his gun; but he saw at once that he could not shoot without killing his brother, and he leaped forward empty-handed. Stumbling over an impediment and falling to his hands and knees, he espied the bloody pocket-knife just dropped by Charley. A moment later the wild-cat was stabbed in the side; then again and yet again.

But poor Charley was still exposed to the wounded animal's cruel claws, and, realizing that he must at once be freed, Joe seized the cat's left fore-leg and pulled with all his might.

The snarling beast was thus partly drawn away from its victim; and Charley's hands, which had gripped its throat, now fell to struggling with its right fore-leg, the claws of which were sunk through his clothing and tearing the flesh of his shoulder.

Then it was that Joe plunged the knife to the hilt in the animal's throat. It was all over after that. Both the windpipe and the jugular vein were probably cut, for in a few moments the cat ceased to struggle.

The battle had been won, but not without its cost. Both boys were bleeding from a number of painful, though not serious, wounds, and their clothing, in places, was literally torn to shreds. As soon as it was all over, Charley sat down in the grass and burst out crying.

" I d-don't w-want to cry, Joe,' he apologized between sobs, " but I c-can't help it."

" Never mind. You jus' cry as much as you want to ; don't be ashamed," said Joe, rather unsteadily, and looking as if — but for his weight of years — he might condescend to cry a little himself.

Having wiped the blood from his face, Joe now proceeded to cut a long green stick. He then fished some twine out of his pocket and tied the wild-cat's feet together. Thrusting the stick between its legs, he took one end of it and Charley the other, and thus they returned in triumph, bearing their dearly bought prize between them.

Whenever anything very unusual or extraordinary occurred, Asa was in the habit of remarking, "Der mus' be a
deviation somewhere;" and when, at sunset that afternoon,
he saw the two boys approaching the camp-fire, all covered
with blood, and carrying a dead wild-cat suspended from a
stick between them, his favorite expression occurred to him
as most applicable. The black man's jaw dropped with astonishment; clearly there was an extraordinary "deviation" somewhere.

A few leaps, and he was beside them; a few words, and
he knew the outline of their story.

"Look yuh, Joe!" he cried, laughing and gesticulating
in an ecstasy, "you don' mean to say you an' Charley kill
dat wile-cat wid des yo' *pocket-knife!*"

"Yes, we did," declared Charley, proudly.

"Oh, go 'way! Well, well, well, ef dat don't beat all!
W'y, you boys — you two boys," the negro cried gleefully,
patting them on the back, "I could turn in an' des hug you
two boys!"

Hardly less enthusiastic were the deserters, most of
whom had now gathered to the camp. Some of them expressed their admiration for the youngsters' pluck in no
mild terms.

"That's the sort o' grit I like to see, boys," said Bubber
Hardy, showing great pleasure. "Never mind, son, I'll
fix it," he said kindly to Charley, who winced on being
patted on one of his wounds.

Bubber then carefully washed and dressed the wounds of
both boys, binding some up with strips of cloth and salving
others, the rest of the men standing and looking on. Even

Sweet Jackson spoke a kind word to them, offering for their use a box of salve which he had made from bears' marrow, and the stingy Lofton produced a flask of whiskey and made both boys swallow a little of it, assuring them that it would lessen the pain.

Everybody seemed determined to make heroes of them, and Joe was so much elated that he forgot the pain. If it was worth so much praise to fight and conquer a wild animal, thought he, how truly glorious it must be to shed one's blood for one's country!

CHAPTER IX.

"BREATHES THERE THE MAN, WITH SOUL SO DEAD?"

AFTER supper the men congregated as usual round a
fire a few paces from the one over which Asa did the
cooking, and, lounging about on the grass and smoking,
they began their nightly pastime of yarn-spinning and
jesting. The two boys, the heroes of the hour, were asked
to re-tell in detail the story of their encounter with the
wild-cat, and were honored, not only with many words of
praise, but by an invitation to drink freely of the corn-
beer, a fermented liquor of the deserters' own brewing,
which heretofore had never been offered them.

Charley soon discovered greater attraction in the company of Asa and Billy at the other fire; but Joe lingered
among the men, laughing and talking with almost the
freedom of one of them, though drinking only in modera-
tion. So elated was he that he not only became ob-
livious of the pain of his wounds, but forgot for the
time that he was a prisoner and that his companions
were deserters.

But the situation assumed its normal proportions before
his mind as soon as one "Mitch" Jenkins, who had first
appeared in the camp late that afternoon, began to speak

of the great difficulty he had had in finding the island and of the recent events of his life before entering the swamp.

"I depend ef this Oke-fe-noke a 't a sight," Jenkins was saying when Joe's attention as drawn to him. "Mis' Jackson"— glancing at Sweet — "put me on the trail and told me 'bout how fur it was, but I thought shore I was lost many a time, and calkilated never to git h-yer. I spent one night in this swamp by my lone self, and I don't want to spend nair 'nother. I dunner what I *did n't* hear nosin' an' trampin' round in them ,oods! I thought to myself, I'd a'most as soon be under fire in battle; though I kin jes' tell you it ain't no fun ,o hear the cannons a-roarin' right at you, and feel t   ,s a-whistlin' round yer years, and see men a-fallin' an dyin' all round you, and the blood a-runnin' like water."

"No, it ain't," assented Bubber Hardy; "but you git sort o' used to it after a while."

"Them edicated fellers out o' the towns and off the big plantations seem able to stand it better 'n we piny-woods fellers,— we 'Crackers,' as they call us," continued Jenkins, with the air of one stating a curious and unaccountable fact. "They'll stand up there and be shot down like I dunner what 'fore they'll run."

"They've got more intrust in the fight than we-all have, — that's the reason," asserted Jackson.

"That ain't all. It's becaze they're fightin' mad all the time, and ready to swear they'll conquer or die. That's what they say we must do. They think if we don't whip the Yankees, the end o' the world is a-comin'. I'm willin'

enough to conquer if hit can be done easy, but I *ain't*
willin' to die.

"I waited a mighty long time 'fore *I* deserted, though,"
continued Jenkins, looking around him with an air of
conscious superiority; "and I reckon I would n't 'a' been
h-yer now if they could 'a' fed me. Them soldiers thar in
Verginy is putty nigh starvin,' you hear me; and what 's
a man to do? If a man 's goin' to fight, he 's got to eat, —
that 's what *I* say. Ain't it so, men?"

As no one seemed disposed to gainsay this point, the
new-comer proceeded to describe in detail the increasing
sufferings of himself and his fellow-soldiers of the tottering Confederacy. He also told how he had run the gaunt-
let, — a feat more easily accomplished in these days of
wide-spread disaster than formerly, — and travelled homeward over hundreds of miles of war-scarred country, everywhere now the scene of great privation and trying strait,
filled with a people aghast at their crowding misfortunes,
but unbroken in spirit, and for the most part as loyal as
ever they were in happier times to the cause which they
had chosen to love and uphold, and which represented
their most solemn convictions.

"Oh, I tell you, people is seein' sights these days,"
declared Jenkins. "The Yankees have got Savannah
and Brunswick and 'most everything else, and the piny
woods round the Oke-fe-noke is plum' full o' refugees, —
old men, fine ladies, and little children, — and some of 'em
ain't hardly got a roof to shelter 'em. Hit 's turrible, —
hit 's plum' turrible."

Unable longer to listen quietly to the account of the

accumulating disasters of the Confederacy, Joe had all at once started up and begun to speak. The fermented beverage that he had imbibed added to his excitement, and perhaps entered as a factor into his carelessness of consequences, although he was by nature of a markedly brave and determined spirit. Starting up in the first place in the energy of an excited attempt to controvert a statement made by one of the men, before either he or his companions quite realized what had taken place, he was standing forth and boldly addressing them.

In school he had been noted for his unusual eloquence, and fondness for declaiming martial and patriotic poems; and although he now often hesitated, repeated himself, and mingled his haphazard quotations from these poems with the commonplaces of boyish phraseology, he spoke with real eloquence, and the deserters listened to him in astonishment and admiration. The boy seemed to forget everything but his intense desire to awaken patriotic enthusiasm in the men around him.

Joe asked — in substance — how his hearers could sit supinely and selfishly in the Okefenokee when every man was needed at the front; when one disaster was following fast upon another; when there was a chance that the lost ground might yet be regained if the faint-hearted and faithless would but repent of their evil way and join forces with the brave ! It was true that the struggle was more and more one against fearful odds, but many a battle had been won against fearful odds. " 'Courage,' " cried the boy, with enthusiasm and in stirring tones, quoting from one of his familiar speeches —

> " ' Courage, therefore, brother men ;
> Courage, — to the fight again ! ' "

And even should they fall in battle, was it not a thousand times better to die gloriously than to live in dishonor worse than death ? They talked of suffering and privation — as if it were not glory itself to endure all this for a noble cause ! Would they sell their birthright for a mess of pottage ? Would they barter the beloved Confederacy for a dinner ? What infamy ! The boy declared he would rather starve a thousand times than desert in the hour of greatest need.

After referring with telling effect to several traitors and renegades of history, and the imperishable dishonor attached to their names, the excited young orator assured the deserters — all in his own boyish way — that their disgraceful and selfish flight to the swamp reminded him of the infamous, perjured Scots, who sold their trusting king to Cromwell for a song. He also compared them to the revengeful Coriolanus, who led the victorious Volscians to the gates of his native city, relenting only at sight of his wife, his old mother, and a train of Roman matrons on their knees and in tears at his feet. Did the selfish and cruel men of the present instance, whose desertion had helped to sap the strength of the armies of the Confederacy, wish to see those refugeeing women and children kneeling and in tears at *their* feet ?

Finally he eloquently recited a portion of the " Lay of the Last Minstrel," on the subject of the love of country, which he knew by heart, beginning —

" Breathes there the man, with soul so dead,"

delivering with especial emphasis and fire the concluding
lines : —

> " Despite those titles, power, and pelf,
> The wretch, concentred all in self,
> Living, shall forfeit fair renown,
> And, doubly dying, shall go down
> To the vile dust, from whence he sprung,
> Unwept, unhonored, and unsung."

Then, quite overcome by the violence of his emotions,
the boy turned and rushed away, exclaiming, "O God,
I wish I were a man!" Throwing himself down on the
grass near the fire where Asa, Billy, and Charley stood
listening, he burst into sobs.

"That boy'll run me crazy," muttered Bubber Hardy,
starting to his feet with a snort and striding away into
the darkness.

"Did you ever hear the like?" asked Lofton, breaking
a dead silence. "It everlas'nly made the cold chills run
up and down my back."

"I depend he'd make a powerful exorter," remarked
the man Thatcher, who had been a lay preacher.

"He'll make a brave cap'n in the army one these
days — if the war holds out long enough; that's what he'll
'make," declared the new-comer Jenkins.

"That beer must 'a' made him half tight, or he would n't
'a' dared," growled Sweet Jackson. "He's gittin' a little
too big for his breeches, and he'd jes' better look out. I
don't aim to stand no sich."

As if in answer to this threat, Bubber Hardy now
appeared on the outskirts of the circle of firelight.

" I jes' want to put you all on notice," he said, in a shaken voice, before receding into the darkness: " if anybody lays his finger on that boy, he 's got me to whip."

Meanwhile Asa and Charley had drawn near the sobbing boy, wondering at what they saw; and even Billy seemed sobered for the moment.

" Nem-mind, Mas' Joe, honey," said the negro, soothingly. " Don' cry. It 'll all come out right. Sho' got to be a deviation some o' dese days. I wish yo' pa an' ma could 'a' heard you givin' it to dem mens dat-a way." The negro added, " I know dey 'd 'a' been proud. An' I gwine tell 'em, too, soon 's ever I git de chance, — I gwine tell 'em eve 'y word ! "

CHAPTER X.

THE "COCK OF THE WALK" IS "HURTED" IN HIS MIND.

" I DON' reckon we better try it to-night," whispered Asa, half an hour later.

Joe now sat up, looking dreamingly into the fire, and several of the deserters were climbing up into the loft to bed.

" Try what ? " asked the boy, absently.

" Try to run off like we made out to do dis mornin'. You boys 'll feel too stiff an' bad, won't you, wid all dem scratches fum dat vigeous wile-cat ? "

" No ; we 'd better not try it to-night," was the answer, and Joe relapsed into his revery.

During the next day, both boys suffered a good deal from their scratches, as Asa had foreseen. Charley allowed Billy to draw him into frolicsome play now and then ; but Joe lay quietly on the grass with closed eyes, or watching Asa.

" Mr. Hardy is hurted in he mind, you see him so," the negro said to the boy, late in the afternoon. " Dis mornin' he went off in one de boats all by hisself, an' dis evenin' he ain't done nothin' but walk aroun' all to hissef lookin' powerful serious."

" What 's the matter with him ? "

"I reckon you gie him too big a dose las' night, — putty nigh mo'n he could swallow. Dat man hurted in he mind, *I* tell you!"

"He's more of a man than any of the others," commented Joe. "It's strange he ever deserted. I know he's not a coward."

The boys felt better next morning, and gladly accepted the invitation given them by Hardy to take a trip with him in his boat. Any sort of change was welcome, especially to Joe, who chafed constantly. Hardy announced to the men at breakfast that he was going to Honey Island, and expected to keep an eye open for a bee-tree. Honey had been found on this island more than once before, it appeared; hence its name.

Asa was ordered to prepare a lunch, and the three were soon ready to start. Sweet Jackson observed their preparations narrowly, and before they got off he called "Bud" Jones, and "Zack" Lofton aside, and urged them to take a second boat and accompany the party.

"I jes' bet Bubber aims to turn them boys loose," he said uneasily. "Hit's more'n he can stand to have that boy Joe around, a-carryin' on and a-exhortin' that-a way, and he wants to git shed o' him."

"I'll bet five dollars that's jes' what he's up to," exclaimed Jones; and Lofton gave expression to the same suspicion.

"I'd like well enough to git shed o' that bigity little chap myself," Sweet continued, "but hit won't begin to do; hit ain't safe. I tell you what you two fellows better do, — you go 'long with Bubber to Honey Island, and keep your eye on them boys."

The precaution was one in which all were equally interested, and the two men readily agreed to go. As he was
poling his bateau off from the shore, Hardy was surprised
to see them coming down the slope, each with a musket in
one hand and a bucket in the other.

"Don't you reckon we better go 'long, Bubber?" asked
Jones, persuasively. "Mebby you'll find a bee-tree, and
we kin holp you cut it and git the honey. We was goin'
over that-a way, anyhow."

"All right," was the brief answer; and the two men
sprang promptly into a second boat.

It was soon quite evident to all, however, that the
"cock of the walk" was displeased. During the long hard
pull of three hours over the boat-road, winding through the
flooded swamp and forest, he did not once speak to the
two men, although the distance between the boats was
never greater than a hundred yards, and often not more
than a few feet. But he spoke now and then to the boys,
pointing out objects likely to interest them.

"Honey Islant ain't as big as our'n," he told them
once, "but the bresh is thicker." He then added with
particular emphasis: "On t'other side from where we'll
land, there's a good trail that leads out o' the swamp, —
over land, too; you don't need no boat. *I* could git out
o' the swamp in half a day by that trail."

Joe wondered how long it would take him and Charley
to reach the outer world by the same path; and it occurred
to him that if Asa could only be with them, the three
might slip away, and make good their escape, while the
deserters were engaged in cutting the bee-tree. He was

also a good deal surprised that Hardy should mention the existence of such a trail, little dreaming that the big deserter, in his present troubled state of mind, would gladly see the two young prisoners make their escape. The boys little knew that their friend even felt disposed to assist them in getting off, provided he could do so without exciting suspicion among the men, and provided they would go and leave Asa behind.

Hardy rightly believed that there were a thousand chances to one against the boys being able to guide a party of soldiers to Deserters' Island, even supposing the soldiers could be spared for such duty in these days of misfortune and disaster ; but the odds were far less great against the ability of the negro to do the like, and, besides, Asa's labor was wanted in the camp.

And so the two boys had been invited to go to Honey Island, and on the way the hint was given them, although the prospects of their successful escape were threatened by the presence of Jones and Lofton.

" Charley," called out the last-named when the island was reached, " pick up that piece o' rope in yer boat and fetch it along ; we 'll need it, mebby."

The boats had been run aground several yards from dry land, and all hands were now wading out, Charley being the last to step into the water, carrying the desired coil of rope.

" I b'lieve I kin go right to one," said Bubber Hardy, as soon as they had struggled through the dense " hammock," and gained the higher level of the island. " When I was huntin' over h-yer, week before last, I seen lots and cords o' bees, and I watched which way they was flyin'. If I 'd 'a' had time, I could 'a' spotted one right then."

No one was surprised, therefore, when, less than an hour later, a bee-tree was found. Pausing under a tall pine, the big deserter turned to his followers, and pointed to an almost continuous stream of bees, quite black against the bright sky, issuing from an unseen hole in the trunk of the tree a few inches above the lowest branch, but more than sixty feet from the ground.

It was now midday, and before attacking the tree, the party sat down on the grass, and ate the lunch which Asa had provided. Then, without any unnecessary waste of time, Jones and Lofton rose and vigorously plied their axes on opposite sides of the tree. Scarcely had the chips begun to fly, when Bubber Hardy suddenly addressed Joe in a lowered voice.

"You boys kin take yer gun and run around for a little hunt while we are cuttin' the tree and getherin' the honey," he said. "Maybe you 'll strike that trail I told you 'bout," he added.

"I 'd rather stay and see you get the honey," said Charley, watching the flying of the chips with great interest.

"No, come along," urged Joe, in a tone the seriousness of which his little brother could not mistake.

"You kin git all the honey you want when you come back," said Hardy, smiling at the little fellow.

Charley yielded, evidently against his own wishes; and, involuntarily snatching up the coil of rope, which he had carried so long, he followed his brother into the bushes.

"This is as good a chance to get away as we 'll ever have," said Joe, as soon as they were out of hearing.

"Get away, — without Asa?" asked Charley, astonished.

"Yes. We 'll have to leave him, — we can't help it. I thought it over while we were coming in the boat, and I made up my mind to try it if there was half a chance. If we hurry down to the other end of the island and find that trail Mr. Hardy spoke about, we may get out of the swamp by night."

"But I hate to leave Asa," said Charley, regretfully.

"Never mind. Wait till I guide a lot of soldiers in here; *then* we 'll get Asa!"

Joe had been pushing ahead while he spoke, followed by the half-reluctant Charley, and he now began to move forward in great haste.

The watchful Jones had not failed to note the disappearance of the boys, and he immediately began to show signs of fatigue, drawing his breath very hard, putting in his strokes more slowly, and finally pausing altogether, with an exclamation indicating that his exhaustion was complete.

"Tired out a'ready?" asked Bubber, contemptuously; and, taking the axe, which was willingly resigned to him, he began to swing it with great vigor and despatch.

This was precisely what the cunning Jones desired, and he lost no time in darting into the bushes on the track of the two boys. Half an hour later, as Joe and Charley hurried forward, leaping over logs and dashing through the crowding under-brush, the former happened to glance in the direction whence they had come, and as he did so distinctly saw a man leap behind a tree.

"It's no use, Charley," he said, stopping short. "Bud Jones is following us. I saw him jump behind a tree."

The boys sat down, panting, on a log, and after a few moments Joe proposed that they go forward more slowly a half-mile further, and then return to the bee-tree, just as if their trip had been a hunt and nothing more.

"I'd fight him before I'd surrender, if he were alone," said the elder boy, fiercely, looking toward the spot where he had seen Jones. "But the first thing he'd do would be to whoop up the others, and it would be useless for me to try to do anything."

Joe swallowed his disappointment and chagrin philosophically, and proceeded to give his attention to the pursuit of game, picking his way through the brush slowly and cautiously. At length he halted and signed to Charley to be quiet, as a crow suddenly cawed and flew out of a tree two or three hundred yards in their front.

"That crow saw something, I'll bet," he whispered knowingly.

And when presently fresh bear-tracks were discovered, he added triumphantly, —

"I told you so!"

The tracks soon led them into what was doubtless the path of an aforetime tornado, the ground being crowded with uprooted trees, which had been thrown across each other at every angle, and lay "heaped in confusion dire." Here the trail was lost, but the boys still cautiously advanced.

At the terminus of a hundred yards, standing on an elevated log and looking forward, Joe became greatly excited at the discovery, not twenty feet away, of a small open space covered with a deep drift of pine needles, in the centre of which were two round depressions or beds, some fifteen inches deep and not less than four feet in diameter. In one of these were two young bears, evidently asleep, the mother being probably out feeding.

Signing to Charley to be very quiet but to come quickly, Joe waited until his little brother stood beside him on the log, and had seen what neither were likely to have the opportunity of seeing again. For, indeed, as the deserters afterwards declared, it was a "find" as remarkable as unexpected.

"Don't shoot 'em!" pleaded Charley, as Joe lifted his gun to take aim. "Let's catch one of 'em alive and take it to Billy. We can tie it with this piece of rope."

"Well, we can try it," assented Joe, determining not to fire unless the attempt at capture failed.

Cautiously they stole down the log and stepped upon the soft carpet of pine-needles, but now, unfortunately, a twig snapped under Charley's foot, and one of the little bears lifted its head and looked around. An instant later cub number one leaped to its feet with a gruff snort and bolted into the bushes, but before number two had followed, Joe was upon him.

Letting his gun fall, the boy leaped forward, alighting astride of the cub's back and grasping its ears with both hands. Uttering a peculiar sound, partaking of both an angry snarl and a terrified whimper, the vigorous little bear tried to jump; but Joe exerted all his strength and successfully held it down, the frantic cub meanwhile tearing up the bed of pine needles with its well-grown and powerful claws, and struggling furiously to get at its captor.

By this time Charley had made a slip-knot, as he was directed, and passed the rope around the animal's neck. Seizing firm hold of the other end of the rope, Joe rose and let the cub go.

"We'd better look out for the old one now," he said warningly.

Released, the cub ran away with great precipitation, dragging the boy after it, along a path which fortunately led out into the more open pine woods, and in the direction of the bee-tree.

"Bring my gun," called Joe, and picking it up, Charley ran after him, trying to keep a sharp look-out for the "old one," as he was warned to do.

As long as the cub ran in the right direction, no effort was made to check its progress; but before a great while it turned off abruptly to the right, and then Joe was forced to exert all his strength in order to drag it after him. Even then his efforts would have been comparatively without result, had not Charley, who proudly covered their retreat, gun in hand, frightened the little bear from behind with a frequent shove of his foot.

Within a few minutes Bubber Hardy had become aware of Jones's absence, and he was not slow to suspect the cause thereof; but he went on cutting without a word, concluding that it would be wiser not to interfere. When Jones reappeared, three quarters of an hour later, offering some trivial excuse for his absence, Hardy concluded that the boys had successfully eluded their pursuer. By this time the tree was down, the hollow had been located, and, protected from the angry bees by the smoke from burning rags, the three men proceeded to cut into the tree and secure the stores of honey.

Jones had followed the boys far enough to become convinced that they were really off on a hunt and would ere long return; but great was the surprise of all when Joe and Charley at length appeared, dragging the young bear after them.

"Well, I'll be switched if that don't beat all!" exclaimed Hardy, dropping a bucket of honey and going to meet the boys.

Before there was time for an explanation, a sound as of a hurrying, bulky body was heard in the brush, out of which captors and captive had just made their appearance, warning all hands to be on their guard.

"It's the old one!" cried Joe, and, surrendering the rope to Charley, he snatched his gun and stood ready, just as a large she-bear dashed into the open, and came toward them, snarling fiercely, and clearly determined to battle for her young. There are few animals more dangerous when at bay, or bearded in the den, and, not daring to trust to Joe's marksmanship, Hardy ran for his own weapons.

Joe fired promptly and with good effect, although trembling with excitement. The load of buck-shot pierced the animal's tough hide between the neck and left shoulder, causing it to halt with a hoarse whine or growl of pain. But only for a moment. With a maddened snort, the bear came on more fiercely than ever, until a bullet from Hardy's rifle entered a vital part, causing the bulky form to roll over on its side in the agonies of death.

As Hardy and Joe ran forward to examine the prize, cries were heard from Jones and Lofton, who were now seen running wildly, pursued by dozens of infuriated bees. In their absorbed interest in the shooting of the bear, the two men had forgotten the necessary manipulation of the burning rags, allowing them to go out, and were now simultaneously attacked by the determined little citizens whose walls had been rudely broken open, and the fruits of whose busy labors were being seized.

After a hot pursuit of a hundred yards or more, the bees returned to their rifled storehouse and the robbers were allowed to escape, not, however, before each had been stung several times.

"I saw you sneakin' along behind and watchin' us," said Joe, contemptuously, to Jones, when later all hands stood looking on as Hardy skinned the bear.

"Who, me? I was lookin' for another bee-tree," was the ready answer.

A chain was brought from the boats; and the captive cub, which had gnawed the rope in two and very nearly effected its escape, was permanently secured therewith. Shortly afterward, laden with many pounds of the choicest steaks, the bear's hide, and two buckets of honey, not to mention the young cub, which the boys forced after them, the party returned to the boats and paddled homeward.

CHAPTER XI.

D URING the return trip, strange as it might seem, Joe and Charley were the most cheerful ones of the party. The faces of Jones and Lofton were by this time swollen almost beyond recognition; and the pain and vexation which they suffered often drove them to swear furiously. Bubber Hardy suffered no physical pain, but his mood was scarcely more agreeable. Asa's perceptions guided him aright when he asserted that the big deserter was "hurted" in his mind.

The two boys, on the other hand, were so delighted over their successful capture of the young bear, and had been so much diverted by the many unusual incidents of the day, that they almost forgot their own captivity, and hardly even regretted their recent failure to escape, especially as Joe felt doubts of the advisability of their making the attempt without Asa.

The party reached Deserters' Island an hour after dark, and great was the sensation around the camp-fire when Joe and Charley appeared with their prize, and the honey, the bear-meat, and the skin were exhibited.

Billy almost danced with delight at sight of the cub, and was soon busying himself preparing something for it to eat.

"Well, I depend if them two boys ain't a sight in this world!" declared the new-comer, Jenkins, with undisguised admiration.

"Rafe and Jim come back this evenin'," announced Sweet Jackson, as the "cock of the walk" appeared, referring to two of the deserters, who had gone out of the swamp a day or two since in quest of meal and salt.

"Did they git the salt?" Hardy asked.

"Yes, and two bushels o' meal."

"We kin do without meal, but we got to have salt."

"And they brung a live gander," put in Jenkins. "What do you say, boys? Less have a gander-pullin' to-morrow," he continued, looking from one to another.

"We ain't got no horses nor no race-track," objected one of the men.

"Oh, we'll jes' swing him up and run round and grab at him on foot. We'll git jes' as much fun out o' it that-a way. I've seen it done when ther' want no horses nor mules nair one on hand."

"Anything for a little fun," seemed to be the consensus of opinion, and a gander-pulling for the morrow was forthwith agreed on.

Before he climbed into the loft that night, Joe sought speech with Asa.

"I'm afraid Charley and I won't be able to keep awake to-night," he said. "But we *must* try it to-morrow night."

"All right, Mas' Joe," assented Asa. "To-morrow night, den. Ef we don't git off scot free, we'll gie 'em some fun ketchin' us, anyhow."

The gander-pulling took place on the following afternoon. During the morning two stout poles about fifteen feet long had been sunk into the ground some six or eight feet apart and a rope swung loosely from the top of one to the other. To this, when the hour arrived, the gander's feet were securely tied, so that the fowl's neck swung within easy reach of a man of average height.

About four o'clock in the afternoon the doomed fowl was hung up, its long neck having first been thoroughly greased. Both operations were violently objected to and jealously watched by Billy, who had already adopted the gander as one of his pets.

All hands having gathered to the spot, the new-comer Jenkins, who seemed to be the leading spirit in this festivity, passed round a hat and took up a collection as a prize for the as yet unknown victor. As nearly every one contributed something, the sum raised was not inconsiderable. Asa, Billy, the two boys, and Bubber Hardy formed a party of spectators, all the other men, eight in number, proposing to enter the contest. When asked why he did not take part, Hardy briefly replied, —

"I ain't a-hankerin' after no such tomfoolery to-day."

Lots having been drawn in order to determine who should have the first trial, the second, the third, and so on, Mitch' Jenkins looked about him, with an air of importance and responsibility, shouting, —

"Gentlemen, is you ready? Let 'er go! Everything's lovely, and the goose hangs high!"

Thereupon Sweet Jackson, who had drawn the first lot, took a position on a line drawn about fifty feet from the two posts, and at a given signal started forward at a rapid run. As he neared the swinging gander, his right hand was thrust upward, and he endeavored to seize the fowl by its neck, but without success, the gander cunningly twisting its head out of reach.

A loud guffaw went up from all sides, as this signal failure to wring the fowl's greased neck was witnessed. Bud Jones, the swelling of whose face had subsided, now ran forward and made the attempt with no better success. Then came the turn of Zack Lofton, whose face still mutely bespoke the revenge taken by the despoiled bees. He succeeded in firmly grasping the gander's neck, and but for the treacherous grease, its head would have accompanied him in his onward rush.

Released, the unhappy fowl swung back and forth, hissing and squawking in an extremely ludicrous and yet pathetic manner, exciting the laughter of the crowd, the pity of Charley, the indignation of Joe, and the tears and angry objections of Billy.

"Quit it! Quit it, I tell you! You-all let my gander alone!" screamed the witless young man again and again, as the contest continued.

Once he ran between the two posts and made efforts to take the fowl down, but retired, whimpering, upon receiving a resounding box on the ear from Sweet.

"It's wicked to torture that poor gander in that way," declared Joe, indignantly. "Why don't they kill it at once?" he asked of Hardy.

"Wouldn't be no fun in that," was the answer.

After all hands had made several trials and the gander's greasy neck had received a number of rude wrenches, the poor fowl held its head less high, ceased to hiss, and squawked more plaintively than ever. The game was easier now, and almost every contestant succeeded in grasping the neck as he ran past; but however firm his grip, the gander's greased head would inevitably slip from his grasp.

At length, after the contest had lasted fully an hour and a half, and the object of this cruel sport had almost ceased to make any outcry whatever, Zack Lofton leaped upward as he ran by and grasped the neck of the fowl near its breast. As his body was carried onward by the force of its momentum, his tightly gripped hand slipped like lightning along the gander's neck, but paused at its head. For one moment the man's body swung from the ground, his whole weight supported by the neck of the still living fowl. It was then that he gave his arm and hand a vigorous twist, and the next moment landed on his feet some distance beyond the posts, carrying the gander's head with him.

"Mr. Lofton gits the prize," cried Jenkins, walking over to the victor and pouring the collection into his hands.

"He did n't git it fair," declared the disappointed Jackson, in loud, angry tones. "Who *can't* wring off a gander's neck if he swings on to it that-a way? I say hit want fair."

"We all had the same chance to do what he did," argued Jenkins, good-humoredly. "The trouble was we could n't keep our grip."

"That 's right," agreed several others.

"And I say hit want no ways fair!" repeated Jackson, in great anger.

Flushed with victory, Lofton did not pause to calculate consequences (for Jackson was a dangerous man) and promptly gave his accuser the lie, which, in local parlance, was equivalent to the "first lick."

Sweet Jackson's face turned livid, and, whipping out a long knife, he leaped toward Lofton. The uplifted blade descended before it could be warded off, and, as the other men rushed in and forcibly separated the enraged combatants, the two boys, looking on with all their eyes, noted a long narrow red streak all across Lofton's forehead and left cheek. An instant later this had expanded an inch in width, and presently the man's whole face was covered with blood.

"Oh, yes, Zack Lofton, see now what you got for pullin' off my gander's head!" cried Billy, triumphantly, dancing about and giggling. "See what you got now! I wish my gander knowed it. I'll bet he does know, too. Anyhow he'll know by and by, and he'll laugh. He'll have a good laugh."

"Shut up your tomfoolery!" commanded Bubber, as he passed the half-witted young man, and proceeded to care for the wounded.

Sweet Jackson was forced away in one direction and Lofton in another, both cursing with great fury, and each vowing that he would take the life of the other.

Meanwhile the two boys and the negro remained immovable in their places, wondering what would happen next, until Billy approached and, cutting down the headless body

of the gander, was about to bear it away.   Then Asa inter-
fered.

"Gim-me dat gander, boy," he said, laughing.   "Quit
yer foolin'.   Quit yer behavishness, I tell you!   We got
to hab dat gander fer dinner to-morrer, you see hit so."

Lofton now lay on his back on the grass, and Bubber
Hardy was on his knees, bending over him and wiping
away the blood.   The cut across the cheek was so deep
that it was found necessary to sew it up, to which opera-
tion Lofton submitted without resistance, but groaned as if
in great pain.   Having done all that seemed necessary or
possible, Hardy assisted the wounded man into the loft,
bade him lie down in his corner, and made him as comfortable as the circumstances would permit.

The "cock of the walk" then sought out Sweet Jackson
and spoke to him with a serious, determined air, after
which that pugnacious individual rapidly cooled down,
ceasing the profane and threatening speeches in which he
had been loudly indulging since the moment he was
dragged away from his foe.

Notwithstanding this violent termination to the festive
gander-pulling, the deserters, with the exception of Bubber
Hardy and the wounded man, were not slow to recover
their wonted spirits, and after a hearty supper they sat
about the fire and joked, laughed, sang corn-shucking songs,
and drank in the greatest possible good-humor.

Asa smiled covertly, and shook his head.   This was a
"deviation" of a kind which by no means pleased him.
Where all this would end was more than he could be sure
of, and he trembled for the future.

"Look yuh, Mas' Joe," he said to the boy, with a comical air, " I want to git away fum dis place 'fo' somebody draw a knife on me an' cut my t'roat."

" Well, let 's make a break to-night," the boy proposed, and the negro agreed.

" Dis a good night to try it," whispered Asa, as Joe was preparing to climb into the loft about ten o'clock. " De mose o' de mens is half drunk, an' dey 'll sleep hard — cep'n hit 's Mr. Lofton. You better look out for *him ;* he 'll lay wake mose all night like ez not. Don't you move a foot tell 'way late 'bout two o'clock, or we 'll ketch de ve'y devil."

CHAPTER XII.

FLIGHT.

CHARLEY was told that escape would be attempted that night; but very soon after they had lain down on their bed of moss in the corner of the loft he fell asleep, leaving the responsibility, as was natural, to his brother and the negro. Not so Joe, who lay awake for hours, listening, waiting, planning.

The watchful boy was soon aware that Bubber Hardy, although probably asleep, was very restless, and would, no doubt, be awakened by the slightest sound. As for Lofton, it seemed doubtful whether he slept at all, for every few minutes he gave utterance to a sigh or groan of pain.

At last Joe began to fear that there was no hope of their being able to escape from the loft at all that night, and in the midst of discouragement sleep overtook him.

When he awoke, all was quiet in the loft, except for the loud snoring of several of the men. Neither the restless Hardy nor the wounded Lofton now made any sound. Joe could not tell why he thought so, but he felt convinced that it was near morning. Lifting himself guardediy upon his knees, he bent over his sleeping brother, and endeavored to rouse him.

"Wake up, Charley!" he whispered, his mouth almost touching the little boy's ear. "Wake up! It's time for us to start."

"Let me alone! What are you pushin' me for?" said Charley, stupidly, and so loud that Joe was terrified, and allowed the boy to relapse into slumber.

Having listened intently for a few moments and hearing no one stirring in the loft, Joe made another effort, and presently had the satisfaction of rousing Charley into complete wakefulness without unnecessary noise.

He then took his little brother's hand, and together they crept along the wall until they stood opposite the hole in the floor. On the way, Joe, who was ahead, stumbled over an outstretched foot, and narrowly escaped falling. The disturbed sleeper turned over, grunted, muttered a few unintelligible words, and all was quiet again.

Just as the boys were preparing to swing themselves down through the opening, not daring to put down the ladder, one of the sleepers stirred noisily, and they heard the voice of Lofton demanding, —

"Who's that?"

Drawing back into the deep shadow, the boys stood silent, holding their very breath. The challenge was repeated, but they made no answer. Then, for perhaps a quarter of an hour, they stood in their tracks, hardly moving a muscle, breathing softly, and fearing that even the violent beating of their hearts would be heard.

Convinced at last that the wounded man had relapsed into slumber, they noiselessly swung themselves down through the opening and dropped to the ground below.

Several dogs, lying asleep beneath the loft, now rose and followed the boys with signs of great cheerfulness, evidently anticipating a night hunt.

Their first object in view was to "turn Asa out," as Charley said. Such phraseology suggests a pen rather than a house, and so indeed the negro's nightly prison was called; but in reality it was a rough shanty of large pine saplings, the door being secured from without by a lean-to formed of a section of a heavy log about twelve feet in length. Having lifted this away and let it down gently on the ground, the door was opened, and Asa came forth, rubbing his eyes, and whispering, —

"I clean give you out, and went to sleep. Hit's mose daylight," he added, "an' we better be gwine quick."

After a hurried consultation it was decided to "cut across" the island and take the trail through the jungle, rather than go upon the prairie in a boat, where daylight would soon discover them to view. Besides, on the prairie they were likely to go astray, but, once on the jungle trail, they were comparatively safe in that respect. Asa wanted to secrete the boats as a blind; but it was now so near morning that the time could not be spared.

"Let's take the dogs," suggested Joe, "so that the deserters can't track us. After we get a good start of five or six miles, we can whip 'em and make 'em go back. We'll be out of the swamp then before they can catch us."

Asa agreed to this, and accordingly the dogs were called softly. The whole pack, five in number, followed willingly, as the two boys and the negro hurried away from the camp. The four miles of the island were covered with the greatest possible speed. Wherever the ground was sufficiently open to permit it they ran, Asa leading Charley by the hand; and even in the brush they pushed forward rapidly, careless of scratched hands and faces and torn clothing.

Faint light streamed through the tree-tops from the whitening sky overhead before they had traversed half the length of the island, and by the time they reached its limit broad daylight surrounded them. The fugitives now observed with considerable concern that the dogs had disappeared, and surmised that they had returned to camp.

"Dey knowed sump'n wrong was up," said Asa, confidently.

They soon found the trail and hurried into the jungle, careless of the mud and water, the thorny brambles, and the possible moccasins, absorbed in their intense desire to escape and the necessity of great haste; for they knew well that within an hour's time the deserters would begin the pursuit.

Asa, who led the way, now paused suddenly; and opening a tin bucket which he carried on his arm, he urgently advised Joe and Charley to help themselves to some of the cold bread and meat therein, and put it into their pockets.

"Gwine to be hard to keep tergedder, when de dogs git at us," he said, — adding, "but if you-all git lost fum me, don't you give up; you keep gwine right on by yo'self tell you git home."

Pressing on with great energy for an hour longer, and not as yet hearing any sounds indicating pursuit, they began to feel more secure; and by and by, at the urgent request of Charley, who was beginning to fag, they sat down on a log, and refreshed themselves with some of the cold food.

" We got to be gwine ! " cried Asa, some fifteen minutes later.

He had sprung to his feet, as the distant baying of dogs fell faintly on his ear.   All knew at once that the deserters were on their trail, and that there was no time to lose.

" Yuh, Charley, you git on my back, an' Mas' Joe, you foller behine me an' do des what you see me do," said Asa.

Catching the little fellow up and putting him astride of his neck, the negro dashed forward over the difficult ground, jumping from tussock to tussock, stepping upon roots and masses of dry moss, and avoiding every bit of soft exposed earth where a track would remain imprinted. Whenever a fallen log ran parallel with their course, he sprang upon it and walked its full length.   Once he made a complete circle, two hundred yards or more in diameter, then, springing forward upon a fallen log several feet beyond the limits of this circle, and directing Joe to do likewise, he pressed forward again over the direct course.

This manœuvre was intended to delay the dogs, and perhaps throw them off the scent ; but before a great while it became evident that it had not succeeded.   For the barking, instead of gradually subsiding in the distance, as they had hoped it would, after a short cessation became more vigorous than before, and unmistakably drew nearer. Ere long it was perfectly clear that only a few minutes could elapse before the dogs would overtake them.

" Will they bite us ? " asked Charley, anxiously.

"No," said Joe; "they know us. What ought we to do?" he continued, looking at Asa, who had come to a halt. "Suppose we shoot them? I could load up and shoot them one by one. I'd *hate* to do it, but we have a right to do it."

Joe carried his gun and Charley his hatchet. The negro had only a butcher-knife, but it was a sharp and dangerous weapon, the blade gleaming brightly where it stuck in his belt.

"Better let me go fer 'em wid dis knife," said Asa, shaking his head. "You shoot dat gun, an' de 'zerters 'll know right whar we is."

Further discussion was cut short by a yelp so close in their rear that all knew the dogs would be upon them in a few moments. Bidding the boys conceal themselves, Asa ran back a few yards over the trail, and took his stand behind a large pine.

As the foremost dog, a big ugly cur, rushed past, the negro leaned over, and with almost incredible quickness, seized the animal's ear with his left hand, and with his right brought the long knife upward across its throat, severing windpipe and jugular vein at a single stroke. With a stifled cry in its throat, the dog rolled over on the ground and lay still, whereupon the four others took to their heels on the backward track with yelps of affright. There was not a bloodhound among them, and they were for the most part the commonest of piny-woods curs.

The three fugitives now hurried onward as before, and for an hour they heard nothing more from the dogs. Finally a subdued, and, as it were, muffled yelp began to be heard at intervals. Asa looked puzzled and several times paused to listen, showing great anxiety when he became convinced that the sounds were drawing nearer. At last he told Joe that he believed the deserters held the dogs in leash, their object being to steal upon the unsuspecting fugitives, who would likely halt to rest in fancied security.

" I bet dey 're comin' like forty," the negro concluded ; " an' ef we ain't mighty spry, dey 'll nab us 'fo' we know it."

" Can't we put the dogs off the scent in some way ? " asked Joe, looking about him.

They were now in a dense thicket of poplars and oaks, gay with the first full leafage of spring ; and a hundred yards ahead the ground sloped downward and was evidently covered for some distance with water.

" I believe we could climb up one of these trees and swing from limb to limb until we got out yonder over that water," proposed Joe, eagerly. " Then we could slip down and wade as far as the water went, then climb up again, and if the trees are still thick enough, go on a good ways. *That* would break the trail."

" You mighty right," assented Asa ; " if only we able to do it. Maybe hit 'll be easy enough for you boys, but hit won't be so easy for me."

" Let 's try it, anyhow," urged Joe, and they at once began preparations.

Charley stuck his hatchet in his belt, and with the help of Asa, and by means of some stout twine found in their pockets, Joe strapped his gun across his back. Asa having disposed of his bucket in a similar way, and all now having their hands and arms free, they began the climb.

The youngest, who was light, active, and an expert tree-climber, led the way. Lifting himself among the larger branches of a spreading poplar, Charley found it comparatively easy to walk out on a lower limb, — while grasping a higher, — until he could lay hold of a stout, interlacing branch, and swing himself safely among the larger arms of a neighboring oak.

Joe was probably sixty pounds heavier than Charley, and found the feat much more difficult. The limb which had borne his brother's weight bent dangerously beneath his own; and when he finally seized a branch of the neighboring tree, he grasped it so near its terminus that he swung halfway down to the ground, and had not the bending branch been one of tough oak, it would probably have given way, and precipitated him to the earth.

Hand over hand the boy swung toward the tree's trunk, and once there he halted to catch his breath and watch Asa. The negro might well hesitate, for he weighed nearly a hundred pounds more than Joe. After a few tentative movements, he saw clearly that his only hope was in a bold leap into the branches of the neighboring tree, trusting to the spreading of his arms and legs, and to his quick, firm grasp to arrest his descent to the ground.

The sound of a muffled yelp from the dogs, unmistakably coming from a point only a few hundred yards away, decided Asa. He took the dangerous leap, and landed among the stout branches of the oak unharmed, save for a few scratches and bruises which he scarcely felt.

CHAPTER XIII.

"WHY don't you come on?" called out Charley, who had rapidly transported himself from the second tree to a third, and from the third to a fourth, imagining, with boyish vanity, that his superior speed was due solely to superior agility.

Joe followed more slowly and warily, but surely. In about ten minutes, the two boys had transported themselves more than a hundred yards without once setting foot on the ground, and were now above the water.

"Don' wait fer me," called out Asa, softly, in answer to a low whistle from Joe. "Git down in dat water an' go on fas' ez you kin. I'll git dere bimeby."

Swinging themselves down from the tree in which they had halted, Joe and Charley waded forward in water varying in depth from one to three feet. At the end of about a hundred and fifty yards the land sloped upward again, and the boys saw what was comparatively dry land ahead of them. However, they were afraid to set foot thereon as yet, and, climbing a tree (for the vegetation was almost as dense where the water stood as elsewhere), they swung themselves forward as before.

Meanwhile Asa was in trouble. After leaping successfully three or four times, he at last — while the boys were

wading forward in the water — failed to gain a firm hold
of the branches through which his heavy body descended,
and, though his fall was broken, he struck the ground with
great force, and was for a few moments considerably
stunned.

A sudden chorus of barks from the rapidly approaching
dogs roused him to action. Struggling to his feet, he laid
hold of the poplar-tree, into which he had attempted to
jump, and climbed with some difficulty into its branches.
The unfortunate man felt that he could not jump again,
and that the only and forlorn resource open to him was
to conceal himself as best he could in the foliage of the
tree.

Scarcely had the trembling of the leaves and branches
subsided, when the pursuers arrived. The party consisted
of Sweet Jackson, Bud Jones, and three others. They
held the dogs in leash, as Asa had suspected, but were
marching with the greatest possible speed. Reaching the
point where the trail came to an end, the dogs one and all
halted, snuffing the air in a mystified way, and could not
be forced forward.

" They must be round h-yer some'rs," declared Sweet
Jackson, in his usual loud, grating voice.

The two boys had halted in the same tree in order to
wait for Asa ; and, on looking back and observing that he
had not even reached the water as yet, they became
alarmed. Joe was about to whistle, when he heard Jackson's familiar voice and a moment later the yelps of the
puzzled dogs.

" Oh, now they 'll catch Asa !" cried Charley.

"Hush!" cautioned Joe; and then the two remained silent, listening intently.

"Mebby they tuck a tree," the boys now heard one of the men suggest.

A silence followed, and it was evident that the members of the party had separated and were scanning the neighboring tree-tops. Suddenly one of the dogs began to bay, and a few moments later, Bud Jones' voice was heard, —

"H-yers one of 'em up this tree!"

The dog had snuffed the spot where he fell on the ground, and poor Asa was discovered. "It's the nigger," added Jones.

"Shoot 'im, if he don't git down from thar quick," cried Jackson, savagely.

Instantly the branches of the poplar began to tremble, and Asa descended with all speed.

"Now whar's them two boys?" demanded several at once, as the negro was roughly seized, and his hands tied behind his back.

"Who me? I dunno w'ere dey is," declared Asa.

A chorus of angry curses greeted this speech.

"My hands jes' *eech* to git a hold o' you," cried the truculent Jackson. "Ef you was n't Bubber's nigger an' he told me not to beat you, I'd break ever' bone in yo' body. *Whur* is them boys?"

"I can't tell you," stammered Asa, determined not to give the boys up if it could possibly be avoided. "All I know is dey's a fur ways fum yuh. Dey got lost fum me 'way back yonder w'ere we fout de dogs."

Ejaculations of incredulity greeted this falsehood, and Asa

was threatened with direful and immediate punishment if
he did not tell the truth; but he stuck to his story and
finally it carried conviction, although his captors beat the
neighboring brush for half an hour, endeavoring in vain to
start the dogs.

"That was Rafe Wheeler's dog, you killed an' I reckon
he'll send you to see old Nick before he's done with you,"
was the last threatening speech addressed to Asa which the
boys overheard; and shortly afterwards they felt convinced,
from the few and faint sounds reaching them, that captors
and captive were marching backward over the trail.

"I'm so sorry they caught Asa; they'll beat him,"
whispered Charley, tears in his eyes.

"Never mind," replied Joe. "Just wait until I guide a
company of soldiers in here, then Asa'll be revenged."

By this time it was high-noon, and being no longer in
fear of immediate capture, the boys had leisure to discover
that they were very tired and hungry. But they well
knew there was no time to be lost; and, as soon as they
had eaten what remained of the cold meat and bread given
them by Asa, they descended from the tree and pushed
forward.

Soon after they had penetrated the jungle that morning
the trail gradually faded, until Asa doubted whether they
had really found it in the first place; and after the dogs
were heard on their track, the negro made no further effort
to follow it, but pushed rapidly ahead in the general direction taken, choosing the most open and passable ground.
This was Joe's plan now.

Toward mid-afternoon the ground began slowly to rise

before them, and the forest growth to become less and less
dense, until finally they emerged from the low jungle
region, and found themselves on an open pine ridge where
the ground was covered with wire-grass and dotted with
clumps of fan-palmettos.   They believed they were now
clear of the swamp; and Charley was in the act of exclaiming in his delight, when Joe stopped him.

"Hush!" said the elder boy, — "look yonder."

He pointed out a large, full-grown black bear about two
hundred yards away.   The animal was engaged in pulling
up young palmetto shoots and eating the sweet and tender
part near the root.   After each pull it would rear up on its
hind-legs and look cautiously over the tops of the palmettos
in every direction.   And so, no sooner had the boys seen the
bear, than the bear saw the boys, and bolted precipitately
into the palmetto brush before Joe had quite levelled his
gun.

"I smell smoke," said Charley, suddenly.

They had now tramped out into the open pine woods
some half a mile, and the wind which blew into their faces
wafted a distinctly smoky odor, suggesting a forest fire.
The probability of this was presently confirmed by the
sight of birds, insects, and here and there an animal, as a
deer, a fox, squirrel, or a skunk, making rapidly toward
the swamp.

"Somebody must be burnin' off the woods for the cattle,"
said Joe.   "If that's what it means, we are certainly out
of the swamp at last."

He referred to the common practice among the settlers of
the backwoods bordering the Okefenokee of firing the

woods in spring, in order to destroy the year's crop of tough brown wire-grass, and so give place to a tender green growth on which the cattle might feed to better advantage.

In a short time the boys began to see the fire here and there, and ere long they were confronted by an unbroken barrier of flame extending across the whole ridge. Their position was becoming every moment more dangerous, and Joe looked about him with some anxiety. The swamp half a mile behind them was a certain refuge, and he calculated that they could easily reach it ahead of the fire, but he was reluctant to turn back. While hesitating, his eye fell upon a small cypress pond, about three hundred yards to the left; and he immediately started toward it on a run, calling on Charley to follow.

Joe felt sure that, even if there were no water in the pond, the fire would not penetrate it. "Pond" is hardly the word to apply to these little groves of several dozen cypresses which are so frequently found in the pine barrens, although they always stand on low, swampy ground, which in wet weather is likely to be covered with a foot or two of water. A small puddle, about twenty feet in diameter, was found in the centre of this one, but the boys did not wade into it. As soon as they stood among the cypress "knees" and trod upon damp ground, they felt safe.

Unlike the banyan-tree, which sends branches downward to take root, the cypress lifts its bulky parasite upward, in the form of what is locally termed a "knee." In the submerged swamps the boatman finds it necessary to look out for these very carefully, as they are often hidden just below the surface, and are as dangerous as unseen rocks.

During a few moments, hot smoke filled the space surrounding the boys and almost stifled them; but the fire itself merely burned round the edges of the pond and then passed on its roaring way, the wind soon clearing the atmosphere. After waiting some little time for the ashes to cool, the boys emerged from their retreat and picked their way across the blackened ground.

The wire-grass had entirely disappeared before the devouring flames, but the pines and scrub-oaks stood for the most part intact. Here and there some fallen and well-seasoned log still burned vigorously, and in a few instances, fire had run up on the oozing sap to the tops of the tallest pines.

Joe and Charley tramped over the blackened and heated earth a distance of about a mile and a half, hoping ere long to discover the shanty of some settler. But when at last they reached a "hammock" growth, and descended through it to the borders of a vast prairie or marsh, in every respect similar to the one adjoining Deserters' Island, this pleasing hope was given over with sighs of regret.

It was now perfectly clear that they were still within the borders of the great Okefenokee, and that they had just traversed one of its many islands or portions of elevated land. The origin of the fire puzzled Joe greatly at first; but he concluded, with reason, that some hunter, or some one of the deserters, had recently been there, and the neglect of this person or persons to put out the camp-fire had resulted in the present extended conflagration.

"It's going to rain," said Joe, suddenly, looking up at the sky now rapidly darkening with clouds; "and we'd better fix some way to camp right away."

A favorable spot on the outskirts of the hammock was chosen, and they hurriedly erected a "brush tent," similar to one or two which they had seen constructed during their stay among the deserters. A slender sapling was cut down and lashed at either end with bear-grass thongs to two trees about ten feet apart. Against this cross-bar, which was about four feet from the ground, eight or ten other saplings were leaned at an angle of about forty-five degrees, and less than a foot apart. Over these were then arranged upwards of one hundred palmetto fans, cut within a few feet of the spot, thus forming a thatch, which was protected against gusts of wind by two or three other saplings laid diagonally across. Such a palmetto-thatched lean-to provides a fairly good shelter, as long as the wind blows at the back against the thatch, and not into the open front.

It was nearly dark when the work was finished, but it had not yet begun to rain. While Charley now gathered wood for their camp-fire, Joe took his gun and stole off into the woods, hoping to kill something for supper. He had scarcely advanced three hundred yards, when he saw a large bird fly through the tree-tops and alight on a branch within easy range.

Joe fired, and the game dropped. When found and brought out into the open pines, where the light had not yet entirely failed, the boy was delighted to discover that he had shot a wild turkey.

Some moss and brush having been gathered, and spread on the ground in the acute angle of the lean-to, and portions of the turkey having been broiled with fair success on glowing coals raked out of the fire, the boys satisfied their hunger, and lay down with a feeling of comfort which seemed hardly in keeping with their continuing misfortunes, and which cannot be said to have been lessened by the harmless patter of the rain-drops on the thatch over their heads.

CHAPTER XIV.

LAND OF THE "TREMBLING EARTH."

IN the early morning they were awakened by the rain falling on their faces, and found their erstwhile dry and cosey retreat now thoroughly wet and uncomfortable. Not only did water percolate through the hastily constructed thatch, but, the wind having changed, the rain now beat in from the front. A slow, steady downfall had evidently continued throughout the night.

"It's a set-in rain, and we're goin' to have a hard time of it," said Joe, ruefully.

It was only with the greatest difficulty and after prolonged effort that they succeeded in building a fire, and by the time the remainder of the turkey, which had been hung out of the reach of marauding animals the night before, had been cooked and eaten, it was late in the morning.

What to do next was the vexing question. Even the night before, Joe had been troubled to answer. He disliked to turn back, fearing a possible encounter with a pursuing party of the deserters; but the prairie barred further progress, unless the boys were willing to take the great risks involved in wading it, — through mud, slime, mosses, rushes, "bonnets," and what not, the water being in many places over their heads.

" Let's try it, Charley," Joe proposed at last. " We are wet to the skin anyhow ; and if we can't wade across, we can come back here, that's all. If we once get across that prairie, I don't think it will take us long to find our way out of the swamp."

The younger boy expressed his willingness to follow wherever the elder might lead, and preparations for the trying trip were at once begun. Both boys were good swimmers; but Joe was too wise to venture on a flooded marsh of unknown depth without some safeguard. As they had no boat, and would probably be unable to float a raft, even if one could be successfully constructed, he decided to take with them a section of a tree, to which they might cling, in case they should advance beyond their depth, and be unable to swim on account of the mosses, etc., with which the marsh water at so many points was crowded.

After considerable search Joe found a dead pine which had broken into parts in its fall before a wind storm. A section of this, about fifteen feet long and a little more than a foot in diameter, was chosen. Having provided Charley and himself with a light slender pole some twelve feet in length, and strapped the gun, hatchet, powder-horn, shot-pouch, etc., between two short up-reaching branches of the log, although this promised to be almost a useless precaution as long as it rained, the boys proceeded, and not without considerable difficulty, to launch what Joe termed their " life-preserver."

While they were accomplishing this task, Charley made his first acquaintance with the great curiosity of the

Okefenokee, which may be seen along the shores of almost all the islands within or bordering the prairies. Stepping off from the island shore, the little boy walked forward upon a seeming continuation of the land, — a mass of floating vegetable forms, intermingled with moss drift and slime, forming a compact floor capable of sustaining his weight, which, although it did not at once break through beneath him, could be seen to sink and rise at every step for twenty feet around.

"Why, this ground moves!" cried Charley, astonished.

"You'd better look out!" cried Joe. "It won't hold you up much longer. It's not ground at all; it's floating moss and stuff—"

The speaker paused suddenly, as Charley now broke through, and stood in mud and water nearly up to his waist.

"The deserters call that moss and stuff 'floating batteries,'" continued Joe. "I don't know where they got such a funny name. Father knew about these places, and *he* said the Indians called them 'trembling earth.' That's what the name of the swamp means, — 'Okefenokee,' or 'trembling earth.'"

Once they had dragged their "life-preserver" over the "floating batteries," or "trembling earth," the boys made better progress, although they still had to contend with a submerged slimy moss of a green color (*sphagnum*) and a great variety of crowding rushes. As they staggered along, dragging the log, now only up to their knees in water, now bogging in the yielding ooze till the water rose above their waists, they were for a time much annoyed by a little black bug haunting the sedge, which stung like a wasp.

The clouds still dropped a slow drizzle, and a mist lay upon the great marsh, in which the many little islands, clothed in dun-colored vegetation, loomed up in dim, uncertain outlines. As he looked toward them, Joe remarked that he had heard the deserters call the islands " houses," but that to him they now rather suggested huge phantom ships.

Many cranes, herons, and " poor-jobs " had already risen at their approach and disappeared in the mist; and as they advanced farther out on the marsh where the water deepened and the sedge began to thin and to be succeeded by " bonnets," large flocks of ducks flew up, and occasionally a curlew skimmed across their course.

Passing within a few hundred yards of one of the little islands, they noted that it was grown up at the edges with low " casina" bushes, and that other vegetation sloped gradually up to two or three tall cypresses in the centre, the whole being drearily decorated with trailing drifts of gray Spanish moss, intensifying the already weird aspect.

" It looks like a big circus tent," said Charley.

The water still deepened ; and ere long they were obliged to swim, Joe with his left arm thrown over the forward end of the log, and Charley with his right resting on the rear end. A few hundred yards further on they entered an open and perceptible current flowing almost at right angles to their course.

" Let's follow this," proposed Joe. " It will be so much easier to carry the log."

So they swam on, floating their log with the gentle current which flowed narrowly between the bordering "bonnets," little dreaming that they were on the head waters of the famed Suwanee River.

Half an hour later they were startled at sight of a large turtle, more than three feet in length, floating lazily on the water as if asleep ; and as it sunk out of sight Joe began to feel some apprehension, recollecting that the deserters had said the prairies were full of " 'gator-holes."

How far they travelled, floating on this current, they hardly knew, being unable to see landmarks for any distance. As soon as one of the ghostly little islands floated past and disappeared in the mist, another would be outlined in their front, and, being so much alike, the effect was very confusing. It was difficult to estimate either the distance they had traversed or the time that had elapsed.

" Oh, I 'm so cold and tired and hungry ! " protested Charley at last, and begged that they might land on the next " house."

Accordingly, as soon as they were opposite another island, Joe struck out toward it through the " bonnets " and sedge, dragging the log after him. In this way they came presently into a little round open pool about a hundred feet in diameter, heedless of several dark floating objects a short distance ahead. Suddenly the water about them became curiously agitated, and with a cry of horror Joe looked toward Charley.

" Jump up on the log ! " he said. " We 're in a 'gator-hole."

Neither of them could afterward have told how they did it; but almost in a moment both stood on the log balancing themselves with their long poles, which were thrust down to the bottom, the water being only about seven feet in depth. Under their weight the log sank so low that it was almost entirely submerged, and the position of the two boys was little improved, supposing they were to be attacked.

The pool now seemed alive with alligators, large and small, for fifty feet around; and the boys were greatly terrified, although the huge scaly creatures still lay quiet on the water or swam lazily about, gazing at the intruders with their black, lustreless eyes.

"They're going to eat us up!" gasped poor Charley, hardly able to maintain his upright position.

"Don't be afraid," said Joe, in a low voice, although desperately afraid himself. "They don't look as if they wanted to hurt us. See how quiet they are."

He then suggested that they pole the log out of its dangerous neighborhood, and this they did very slowly and cautiously, lifting their long sticks halfway out of the water and guardedly thrusting them to the bottom again. Although they passed within a few inches of some of the reptiles in the course of their retreat, the latter were not roused from their sleepy indifference, and permitted an easy prey to escape them.

There are doubtless many thrilling alligator stories which are vouched for on "good authority;" but it is a fact that the species found in Southern Georgia and Florida have been rarely known to attack man except in self-defence.

Shortly after leaving the alligator-hole the boys entered shallower water, and soon waded to the shores of the little "house" or island. Leaving their log safely lodged on the "trembling earth" formation, and having struggled through and over the same, they landed on firm but damp and spongy ground. The island was circular in form and hardly two hundred yards in diameter. Casina bushes fringed the shores, the vegetation gradually rising thence to a few tall cypresses in the centre. Everywhere the funereal Spanish moss fluttered in the breeze.

It had now ceased raining ; but a dense mist still floated upon the great marsh. The raw atmosphere was no less cold than the water had been ; and the boys moved about shivering and most forlorn, bitterly regretting their attempt to cross the prairie. The wildness and desolation of the scene was in a manner intensified by the presence of two small gray eagles, which screamed in a harsh, shrill way, as they hovered about a large nest in the top of the only pine-tree on the island.

The extreme weariness of their bodies and their sharp hunger were the only certain indications of the flight of time ; but as the light began to wane, the boys realized that they had been on the marsh many hours and had not landed on the island till late in the afternoon.

It was now necessary to make some preparation for the night, and that speedily. An attempt to build a fire had failed completely, the wet matches refusing even to ignite, and as the gun and ammunition were also wet, there was no hope of obtaining even the raw flesh of a bird for supper, supposing they could have eaten it.

9

"If we only had a fire," sobbed Charley, shivering, "I would n't mind being hungry."

The poor little fellow's distress was presently further increased by dread of snakes. As Joe moved about making preparations for the night, he very nearly stepped on a large moccasin, which he succeeded in killing with a long stick. It was an unusually large one, probably measuring not less than eight inches around the middle, and doubtless the mother of a numerous progeny.

Joe had often heard the deserters describe how they made shift for the night when caught out on the prairie or on a damp tussock in the flooded forests, and he now proceeded to strip bark off the cypress-trees with the aid of Charley's hatchet. This was spread on the wet ground to lie upon, and a quantity of the Spanish moss was gathered to cover with. The latter was damp,—in fact, water-soaked; but even so they would be warmer covered with it than if they lay exposed to the currents of raw air.

These preparations were hardly complete when it began to grow dark. Joe thought they ought to remain awake and keep their bodies in something of motion all night, in order to prevent taking severe colds, but they were both too weary to persevere in such efforts. Sitting on the cypress bark and leaning their backs against a tree, the wet moss drawn up over their bodies, they soon subsided into quiet of limb and tongue, and after a time fell into troubled, dream-haunted slumber.

"We'll never get home," sobbed Charley, while still they talked. "We'll starve to death in this swamp."

Joe made no reply at once. He was thinking how different had been the experience of Robinson Crusoe and other heroes of romance who had been wrecked on unknown islands or lost in desolate places. None of these, he thought, had ever suffered such continuing miseries of body and mind as were now his and Charley's portion. The hardships suffered by such as Robinson Crusoe were indeed severe; but there seemed to be always a wreck at hand with plenty of good things on board to eat, and the castaways could at least manage to sleep warm and dry.

"I hope not, Charley," responded the elder boy, cutting off this train of thought; "but if we do starve to death, it will be all for the best, as father would say."

Joe was perhaps never more acutely miserable in his life than now, and had he been alone his soul could scarcely have risen above the trying surroundings; but the consciousness that his little brother, one weaker than himself, was suffering as much, perhaps more, than he, roused in him a manly fortitude.

"We'll come out all right in a few days," he said, with forced cheerfulness. "But if we don't — well," he added, solemnly, "this world is not everything. If we have to leave this one, we go right into another one; we can never really *die*. That's what father says, and he knows. He said Socrates said, 'no evil can befall a good man, whether he be alive or dead.' That means, if we are honest and truthful and manly, and never want to harm anybody, we're all right, whatever happens. But if we are mean and selfish and untruthful, and love to injure other people, all the riches in the world can't help us or make us *men*.

"I once heard father say," Joe continued, "that every

misfortune will in some way at last really be a blessing to the sufferer; and I thought how wonderful that was. Father said misfortune had made him a better man. And he told me once that it did n't matter so much what we were in the world, whether rich or poor, or what happened to us; what *did* matter was whether we always thought and intended to be, and were, honorable and just. He said that was the great thing. If we do that, nothing can hurt us."

" Even if the alligators or a panther was to eat us ? ' asked Charley.

" Even then," was the answer with a shudder.

" But, oh, Joe, it would *hurt!*

" Not as much as you think. Father said if I were to fall over a high precipice I would become unconscious before I struck the ground, and I think it would be just that way if a panther were to jump on you; before he really began to eat you, you would n't know anything."

Whether it was the result of this comforting philosophy or of sheer physical exhaustion was not clear ; but Charley presently became quiet and soon after fell asleep.

CHAPTER XV.

JOE, however, remained awake a long while listening to a curious recurring sound out on the marsh, suggesting the harsh clank of two pieces of sheet-iron when precipitated the one against the other, which, as he learned afterward, was made by sand-hill cranes when frightened and forced to shift their positions. The wakeful boy could not account for it, and it added no little to the misery of his situation. Another occasional sound troubled him less,— a hoarse bellowing which he supposed to come from the alligators. When at last he slept, it was only to dream of moccasins and alligators, and a nameless, shapeless monster out on the marsh with a metallic gong in its throat.

As the first gray light of morning struggled through the mist still enveloping the marsh, Joe started up and looked about him. His attention was at once attracted to a white sand-hill crane fully five feet in height standing on a point of the island about fifty yards distant.

Seizing his long stick, the boy crept toward the fowl behind the screen offered by the casina bushes. He hoped to knock it down, conjecturing that even the fishy flesh of a crane would be palatable to one half starved. But the wary bird spread its wings and flew away in the mist long before Joe was near enough to use his weapon.

The boys both found themselves suffering with sore throat and their limbs felt cramped and numb; but they were a good deal rested and their desire for food was less active than the night before. On the whole, they felt better and were eager to go forward and try to improve their condition. Joe remarked that if he could only see the island they had left the day before, he would "go right back" there; but if they attempted it in the fog, a thousand chances to one they would go astray, and he thought they had better take the risk of pushing forward.

So they struggled through the "trembling" and breaking "earth" surrounding the island, got their log afloat, pushed out into the little stream, and swam with the current as on the day before. Although their exertions soon began to tell on them, weak for the want of food as they were, the boys pushed forward heroically during the greater part of the day, landing two or three times on the dreary and inhospitable "houses."

One incident of importance occurring on that trying day may be mentioned. Toward noon, while swimming with one arm over the end of the log, Charley's feet and legs became entangled in the rushes; and, losing his hold, he was drawn beneath the water just as a faint cry escaped him. Joe looked back in time to see him go down, and, swimming to his aid, succeeded after considerable difficulty in extricating him, though not until he had swallowed several gulps of water and was pretty badly strangled.

Meanwhile the log had floated with the current, and lodged among the "bonnets" about two hundred yards down stream, and this distance Joe was obliged to swim without artificial aid, supporting his helpless little brother. The last few yards was the scene of a desperate struggle to keep above water until the log could be grasped. After this the boys were forced to land and rest on the nearest island, which fortunately was not far away.

That night was spent, like the preceding, on a "house," and, if possible, was yet more uncomfortable. They were again unable to start a fire, and lay down as before on cypress bark, covered with the damp moss. The pangs of hunger were now extremely painful; and though he made a brave effort, Joe found himself unable to take the same comfort in his father's philosophy as on the previous night, or to soothe poor little Charley with as much success. But he could at least express tenderness and sympathy, and he held his sobbing brother tightly in his arms for a long while.

"Never mind, darling, never mind!" he whispered again and again, — a demonstration of affection of which, perhaps, he would have been ashamed in happier times.

The morning of the third day dawned bright and clear. Not a vestige of the fog was to be seen anywhere on the great marsh. Although they now felt weak and ill, their eyes ran water, and their heads throbbed with fever and headache, the boys felt cheered by this change. In every direction but one they were unable to see anything but an expanse of marsh dotted with "houses;" but in that one direction they clearly discerned, not more than two or three miles away, a wall of green pines, indicating the presence of a large island or mainland. With great delight they noted also that the intervening marsh, though

covered with water in places, was not of a character to
necessitate swimming.

Lifted high with hope, they started eagerly in the direction of the green wall of pines, soon finding, however, that
it was no child's play to cross this portion of the marsh,
scantily covered with water though it might be. For it
was in great part a treacherous quagmire, and the boys
sometimes sank down suddenly in the mud to their armpits. Once Charley bogged up to his neck, and nothing
but his long stick saved him. They had left their log
behind, but fortunately carried their long poles.

It was near noon when they at length reached the high
land where the pine-trees grew. After plunging into a
neighboring pool of comparatively clear water in order to
wash the mud and slime from their bodies and clothing,
the boys climbed wearily up the slope and lay down in
the warm sunshine, shading their faces with palmetto
leaves.

Here they rested two or three hours, then pushed forward wearily but determinedly across the island, if island
it were. The vegetation was soon found to be unusually
dense and wild. Even after gaining the crest of the slope,
where, on the other islands a comparatively open pine ridge
was usually found, they were confronted by the wild brambles of the jungle and immense thickets of scrub-oak and
blackjack.

About an hour later, however, they emerged upon an
open pine barren, where the underbrush consisted solely of
the ubiquitous tyty, hemleaf, and fan-palmetto. It was
here that a herd of cattle was discovered, and the boys

were led thereby to believe that they were now at last clear of the vast Okefenokee.

Great was their surprise, therefore, to see that as soon as their approach was observed, the herd took fright and fled wildly into the brush, only an immense bull standing his ground, facing the boys, head down, and pawing the earth in a threatening manner.

"Why, they must be *wild* cattle!" Joe exclaimed.

The words were scarcely uttered, when with an angry bellow the bull charged at full speed.

"Run, Charley! Climb a tree!" cried Joe, standing his ground for a few moments in order to draw the pursuit upon himself.

Seeing that his brother was almost if not quite out of harm's way, Joe, too, turned and fled, the bull close at his heels. Dropping his gun, he leaped upward and caught the limb of a scrub-oak, and swung himself up out of reach just as the maddened animal dashed past with lowered horns.

Wheeling about, the great bull charged the tree, butting it with great fury. Although it was slender, and trembled and swayed at every shock, the young oak was tough, and withstood the attack, until baffled Taurus had exhausted his rage or his powers, and, retiring, trotted off into the brush on the track of the herd.

As soon as it was safe to venture from their retreats, Joe called to Charley, who was in a neighboring tree, and they descended to the ground.

It was now past four o'clock in the afternoon; but they still pushed on, until Charley fell rather than sat upon the

grass, declaring that he could go no further. The last
mile had been for him literally a drag. All the nerves
of his weakened frame were throbbing with fever and
excitement.

" I feel as if my head would burst," he said, staring
stupidly about him.

Joe, who felt little better, sat by him a while, and tried
to encourage him.

" You stay here and rest, Charley," said the elder boy at
length, rising to his feet, " while I look around for a good
place to camp. The matches are dry now," he added,
" and I think we can have a fire to-night."

An hour later, as the sun sank out of sight behind the
woods, Joe, who had chanced upon something like a trail,
and followed it for a mile, stole guardedly through an oak
thicket, and, halting on its borders, looked into an open
space where a camp-fire burned.

Everywhere in the little clearing there were evidences
of a long sojourn. The stumps of several trees showed
that the felling had been done months, perhaps a year or
more, before. Curing hides hung against the trees ; tools
and cooking utensils lay about on the grass. A pot swung
over the fire from a tripod of three long sticks, and in it
there evidently simmered a savory stew. No dog was
aroused by Joe's approach ; and the sole human occupant
of the clearing was a white man of middle size, with long
iron-gray hair and beard, who sat on the ground near the
fire, his back to the observer.

What he was doing, Joe could not see, and did not wait
to ascertain. After one swift glance, the boy quietly

retraced his steps through the thicket, and ran backward over the trail with all speed toward the spot where he had left his little brother.

"Oh, Charley!" he cried, as soon as he was within speaking distance, "I 've found a camp, and there 's a man there cooking supper!"

But Charley only looked at his brother stupidly, and spoke of his head. Apparently the fever had entered his brain. A great fear fell upon Joe; and although he was now scarcely able to drag one foot after the other, with sudden resolution he lifted the unresisting boy in his arms and staggered along the trail toward the stranger's camp.

> "— Whate'er ye are,
> That in this desert inaccessible,
> Under the shade of melancholy boughs,
> Lose and neglect the creeping hours of time,"

cried Orlando, as he entered the camp of the exiles in the forest of Arden, half carrying his aged footsore and faint-ing servitor.

> "If ever you have looked on better days;
> If ever been where bells have knoll'd to church;
> If ever sat at any good man's feast;
> If ever from your eyelids wiped a tear,
> And know what 't is to pity and be pitied;
> Let gentleness my strong enforcement be."

Joe's appeal, as he staggered into the graybeard's camp, still carrying Charley in his arms, was less stately and picturesque, but doubtless more effective to the startled ear which listened to it.

"Help us — have pity on us," he panted, — "or my little brother will die!"

The boy sank down by the fire with his burden, in a state of absolute exhaustion, as the man with the gray beard started up in manifest affright, and drew back. Evidently he was somewhat deaf, and had not heard the sound of Joe's approaching footsteps.

"Who're you?" he demanded suspiciously, looking around as if expecting some further invasion of the privacy of his camp. "Whur — whur in the *dickens* did you come from?"

Joe did not answer; he lay passively on the ground beside his brother, keeping his eye fixed on the strange man. The question was repeated; and as there was again no answer, the strange man drew nearer, bent over the two boys, and looked at them curiously.

"Are you sick?" he asked more gently.

"Starving," answered Joe, hardly above a whisper.

A wave of compassion swept over the man. He almost leaped to the fire; and, quickly dipping something from the pot into a tin cup, he blew his breath upon it several times, in order to cool it, then ran back to the prostrate boys, and, kneeling beside them, offered the cup to Joe. But the boy gently pushed it away, and motioned toward his brother, indicating that Charley was in the greatest need and should be attended to first.

CHAPTER XVI.

GEORGE WASHINGTON JEFFERSON JACKSON SMITH.

HAVING partaken of the nourishment which was presently offered him again, Joe fell asleep, or fainted, — he could not afterwards tell which, — and there followed a blank. When he again opened his eyes and looked about him, he lay on a bed of moss in a curious circular room, in the centre of which there rose from floor to ceiling what was unquestionably the trunk of a living tree.

Raising himself on his arm and staring about him, no little alarmed to find that Charley was absent, Joe felt the whole room tremble slightly, and heard a sound as of some one ascending a ladder. In a few moments a small slide-door was pushed aside, and the strange man of the long gray hair and beard entered the room. A cheerful expression overspread his naturally kindly face, as he met the boy's eye.

"You feel better now, I reckon?" he said, seating himself on a pile of moss near Joe's bed.

"Where am I?" asked the boy, uneasily, without answer to the inquiry.

"In my house," was the reassuring reply. "You've been pretty bad off, — sort o' wanderin' in yer mind. But you're all right now."

" Where 's my brother ? "

" He 's outside.   He got up and went down this mornin'.
He 's all right.   He jes' had a little fever from cold and
exposure.   You was the sickest of the two.   You 've been
on a harder strain, I reckon."

" How long have I been here ? "

" Three days.   I thought at first you was in for a set-in
spell o' typhoid ; but I reckon it was jes' a narvous fever,
brought on by starvation and so much exposure.   It was
mighty high, though, for a while.   Yer little brother
Charley tole me how you-all 's been lost and a-wanderin'
so long in the swamp.   You boys has seen sights, I tell
you."

" Are we out of the swamp at last ? " asked Joe, eagerly.

" No, not by a long jump.   You 're on Blackjack, one
o' the biggest islands."

Joe heaved a heavy sigh of disappointment, then asked
suddenly, " Are you a deserter ? "

" Who, me ? " ejaculated the man, starting perceptibly,
and turning upon the boy an injured look.   " You don 't
know me," he continued impressively.   " *My* name is
George Washington Jefferson Jackson Smith, and I 'm a
soldier."

" Oh, I beg your pardon, " Joe hastened to say, showing
great regret.   " There are so many deserters in the swamp,
you know, it 's the first thing I thought of.   But," he continued, " where is your uniform, and why are you here ? "

It seemed strange to the boy that this much-denomi-
nated Mr. Smith should appear to be made uneasy by
this question.

"Well, you see," was the stammering reply, "I — I 'm in disguise at present. You must n't tell it, but I — I 'm in h-yer keepin' my eye on that cussed gang o' deserters, an' when the — the — right time comes I — I — aim to bring the soldiers in, and we 'll nab every last one of 'em."

"Oh, will you? I 'm so glad!" cried Joe.

Mr. George Washington Jefferson Jackson Smith now rose and retired, telling the boy he must lie quiet till the morrow. As it was now late in the afternoon, this would not be a very trying task, and Joe willingly acquiesced. Charley's voice was now heard as he climbed up the ladder. In a few moments he entered the room with a smile on his face, whereat Joe was so overcome with joy that he seized his unresisting brother in his arms, and kissed him.

"We are safe at last," he said, and lay back wearily and dreamily on the moss, taking little note of Charley's answering remark, —

"That Mr. Smith is such a funny man. He talks so funny. And he looks just like a ram-goat, with that long beard growin' down in a point."

An hour and a half later tears of gratitude filled Joe's eyes when his host brought in a delicious quail stew for his supper.

"Then you won't want to keep us prisoners," said the boy as he ate, "and won't be afraid for us to leave the swamp, if you 're a soldier?"

"Who, me? No, sir-ree!"

"And maybe you 'd be willing to show us the way out, — you 've been so good to us," continued Joe, with an eloquent look.

" W-e-ll, hardly," hesitated Mr. G. W. J. J. Smith; " I could n't git off for that. You see I could n't spare the time; I 've got to watch them deserters. But I kin put you on the trail. You kin go it by water to the Cow House in half a day. I kin loan you a bateau, — I 've got two, — and you kin leave it for me at the Cow House."

" Oh, thank you! But what is the ' Cow House ' ? "

" It 's a big peninsula runnin' in the swamp. They call it the Cow House 'cause the cattle thieves use' to drive herds o' cattle in there and keep 'em till they could slip off with 'em to some market. That 's whur these wild cattle on Blackjack come from. They run off from the Cow House into the marsh, and come over h-yer and run wild."

" And after we get to the Cow House? " questioned Joe.

" All you got to do is jes' to follow the trail 'bout ten miles through the piny woods, and you 're right *at* Trader's Hill."

Joe's delight at this news was unbounded. He earnestly thanked their new friend, and expressed the hope that some day he might be able to make a fitting return for so many kindnesses.

" Maybe you kin; maybe you kin," was the answer.

It was long before Joe fell asleep, his mind being so full of thoughts of his home, which now seemed so near. In the morning he rose early, feeling well and strong again, and followed Charley down the ladder to the camp-fire. He looked back with great interest at the house in the tree, and spoke of it with admiration to their host, who was cooking breakfast, and who smiled proudly as he remarked that the building had cost him many a day's hard labor. The house, which consisted of one large circular room, was built in a stout water-oak, the upper branches of which had been mostly cut away, the lower serving to support the framework of rough puncheon planks. It had been built in the tree, like the elevated loft of the deserters, as a safeguard against rattlesnakes and moccasins.

" Everything looks as if you had been here a long time," said Joe, glancing about him. The house itself must have stood in the tree a year at the least.

" Ye-yes," stammered Mr. Smith ; " but I hain't been h-yer so *very* long, though. You see ther' was a feller h-yer before me."

Charley now called Joe's attention to two large fox squirrels, lying on the grass near the fire.

" I shot 'em this mornin' 'fore you waked up," said their host. " The woods is chock full of 'em. "

The boys ate a hearty breakfast, after which Joe felt so far restored that he eagerly asked if they could not start for home at once, and only reluctantly yielded when he was advised to rest until the following morning.

The day was spent in talking with their new friend, in giving him some help toward the preparation of the meals, and in lying about on the grass and sleeping. Joe also cleaned his gun, dried his powder and caps, and otherwise prepared for the start on the following morning. Charley took great interest in a bow, belonging to and manu- factured by their host, and considered himself highly honored on being allowed to shoot away two or three

10

arrows, which latter he diligently searched for and returned to their owner. Both bow and arrows were made of ash, the latter being tipped with sharpened bits of steel. The bow-string was made from the tough gut of the wild-cat.

"Come go with me now, if you want to see some fun," said Mr. Smith, at sundown.

He then took bow and arrows, and led the boys about a quarter of a mile away in the woods, telling them he would show them how partridges roosted at night. When the place was reached, twilight had fallen ; but the boys distinctly saw, when pointed out, several birds squatting on a limb of a tree about thirty feet distant.

"Watch me drop 'em," said their host ; and, lifting his bow, he bent it almost double, the string twanged, and the arrow sped on its way.

One of the birds at once disappeared from view ; the others looked startled, lifting and turning their heads from side to side, as if striving in vain to pierce the gathering gloom. Four times the bowman sent an arrow flying, then ran forward himself, and, after a short delay, returned with four birds, each with its head cut off clean. [1]

"Well, *you* are a fine shot!" cried Joe, with great admiration.

"You see, I shoots 'em in the head to keep from spilin' the meat," was the explanation, with a proud smile.

When they had returned to the light of the camp-fire, and their friend was preparing to open the birds, he discovered a folded paper beneath the wing of one of them and called Joe's attention to it.

[1] An occurrence actually witnessed in the Okefenokee.

"Well, well!" exclaimed the boy, having eagerly seized the folded paper and opened it, "that's the very partridge we tried to send a letter to father by. Just think of it!" He then described the circumstances of sending the letter.

"How long ago was that?"

"About two weeks."

"Hit's a wonder that partridge ain't got shed of it long before this," remarked their host. "Birds has got more sense 'n you give 'em credit for."

"Asa *said* that partridge would never leave the swamp," put in Charley.

Joe handed the letter to their friend, intimating that he might read it if he cared to take the trouble. The namesake of the fathers of the republic seemed curiously embarrassed, and, after holding the letter in his hand for a few moments, returned it, saying, "You better read it while I 'tend to these birds;" and Joe did as was recommended.

As they sat about the fire after supper, the subject of the war came up, and their host, as Joe declared afterward, literally "spread himself," becoming very communicative in regard to his own personal experiences. He showed an intense interest in the subject, and expressed unqualified disapproval of the conduct of the war from the beginning.

"Yes, things is goin' wrong," he said, in rejoinder to a regretful remark from Joe. "The truth is," continued Mr. Smith, lifting his index finger into the air in order to emphasize his words, — "the truth is, the war ain't been run right from the start. It never *is* been run to suit me. As I says to Gen'l Johnson after the first battle of Manassas, s' I, ' General, it want done right,' s' I. ' To be shore, it was a victory,' s' I ; ' but it mout 'a' been a long-sight more. If you 'd only 'a' followed 'em up when they run, you mout 'a' tuck Washington,' s' I. S' 'e, ' George, you 're right, as you always is,' s' 'e, ' and I wish mightily you 'd 'a' been on hand to suggest it.' "

" Why — why, what position did you hold ? " gasped Joe, amazed that any private, as he supposed his host to have been, would have dared to speak thus to a general.

" Who, me ? Oh, I was on the general's staff in them days. But unluckily he sont me off that time, and I was n't on hand to tell him what to do."

For a moment Joe wondered how an uneducated and ungrammatical man, such as he saw his host to be, could have found a place on the staff of a leading general ; but the boy was so elated at the thought of being in the society of so great a man and soldier that he did not pause for sober reflection.

" Hit was jes' the same thing at the battle o' Gettysburg," continued this great soldier, with an air of disgust. " The thing want worked right from the start, and I tole Gen'l Lee so myself. I says, s' I, ' General, this won't do — this won't begin to do,' s' I. And the general says, s' 'e, ' George, I done my best,' s' 'e. ' I 'm mighty sorry you want h-yer to holp us out,' s' 'e. ' If I *hed* 'a' been h-yer,' s' I, ' that battle would 'a' ended diffunt,' s' I, for I was rale mad. ' Mebby so,' says 'e, lookin' mighty down in the mouth."

" And you are as well acquainted with General Lee as *that !* " exclaimed Joe, lost in wonder.

"Who, me? I knowed him like a brother. I knowed 'em all. Ther' want a general in the army but what was glad to git my advice. Even the rank and file o' the soldiers knowed me by sight, and when they seen me makin' for the general's quarters, they'd fling up ther' hats and holler, 'Hurrah for George Washington Jefferson Jackson Smith! Make way for George Washington Jefferson Jackson Smith! He's goin' to take counsel with the general. Make way there!'"

"But — but," stammered Joe, still credulous, but struggling in a maze of contradictions, "but why did you come away if they needed your advice?"

"I was mad, for one thing," was the glib answer. "I was plum' put out by the way things was goin', and then, you know, the deserters had to be looked after. The army ain't got no men to lose nowadays. I'm *de*tailed to look after these cussed deserters in this swamp."

"Oh, yes, I see," ejaculated Joe, evidently rescued from further troubling doubt.

"You know, I was named after George Washington and Thomas Jefferson and Andrew Jackson," pursued the great soldier, proudly; "and them men in the army, the generals and the rest o' 'em, use' to say I had the heads o' all three on my one pair o' shoulders."

So their garrulous host went on spinning yarns until a late hour. Finally, Charley, who had fallen asleep, was roused, and then all three retired to their beds of moss in the tree-house, Joe to dream of bloody battles, famous generals, and the society of great men generally. The two boys climbed up first; and as they lay down on the moss, and the ladder was heard to tremble beneath the weight of their host, Charley irreverently whispered to his brother, —

"He makes me think of a ram-goat all the time."

"Hush! you must n't be so disrespectful," said Joe, sternly resenting such levity as an insult to the majesty of that great soldier, George Washington Jefferson Jackson Smith.

CHAPTER XVII.

AGAIN IN DURANCE VILE.

THE boys were well pleased the next morning when their distinguished friend proposed to accompany them a part of the way to the Cow House.

" I want a bait o' fish," he told them at breakfast; "and I think I'll jes' git in t' other bateau and go with you-all as far as the lake."

All their preparations were complete at an early hour, and a start was made. The boys were led about a mile through the woods to a point of the island opposite that on which they had landed. Here two small bateaux were found, and the party embarked on the flooded marsh, following a distinctly marked boat-trail through the water-mosses and grasses.

Two hours later, the boats entered a broad circular expanse of open water, fully a mile across, and passed what might, without great inaccuracy, be termed a shoal of alligators, for the heads of the amphibian monsters could scarcely be counted. They showed neither fear of the boats nor a desire to attack them, but the great soldier prudently made a détour in order to avoid them.

" Soon 's we git past this 'gator-hole," he said, " I 'll show you boys how to ketch a trout."

Once well out into the lake, he allowed his boat to drift, and began to play a "spoon" attached by a three-foot line to the end of his rod. In the course of an hour he had not less than a dozen "rises," and landed safely in the boat four unusually fine black bass, the largest weighing at least eight pounds.

"Take that along for yer dinner," said the soldier-fisherman, pitching one of them into Joe's boat.

Arrived at the opposite end of the lake, he pointed out a boat-trail leading away through the sedge and over the water mosses as before, informing the boys that it would take them "right straight to the Cow House."

"You ought to git there by two o'clock," he added. "Well, good-by, boys; take care yerself."

"You have been *very* kind to us," said Joe, gratefully, as the boats separated, "and I hope we can return it some day."

"Well, who knows but what you kin? You jes' tell yo' pa all about it, and maybe I'll call on him for a favor one these times. Don't fergit to tell him!" shouted Mr. Smith; and after this speech, which struck Joe as being unworthy the man who uttered it, the great soldier waved his uplifted paddle in farewell, and was gone.

The trail was found to be quite distinct all the way, and it was not so difficult to paddle and pole the bateau over it but that they could make fairly rapid headway and might have reached the peninsula at the calculated time. But Joe now felt so sure of reaching home early the next day at the latest that he allowed himself to be distracted and delayed by the game encountered along the route.

Having been unable to shoot his gun for four or five days, the temptation to indulge in his favorite sport was more than he could now resist.

He fired a number of shots at the ducks and other wild fowl rising from the marsh at their approach. Once two wild geese flew over their boat well within range, and after firing, Joe was made happy by seeing one of them plunge, wheel back and forth, and finally fall into the sedge, some two hundred yards to the right of their course. The boy had never shot a wild-goose before and considered it a great prize.

Charley wanted to push on, but Joe would not consent to leave the game behind. Much time was therefore wasted in running the boat out of the beaten track, and in poling it back and forth through the sedge in search of the goose. Several times they ran aground and found great difficulty in extricating themselves. Indeed, the boy was finally obliged to take off his clothes and search for the game on foot, and after securing it, drag the boat back to the trail.

Nearly two hours were lost in this way, and when the boys finally landed on the Cow House peninsula, which seemed in all respects similar to the islands they had visited, it was past four o'clock in the afternoon. By this time they were ravenously hungry, and were obliged to consume another hour in building a fire and cooking something to eat. So it came about that when night overtook them they were still in the heart of the Cow House.

They had selected a suitable spot for a camp, and were building a fire, when the sound of hurrying footsteps caused both boys to start up and look about them. A moment later a man leaped into the circle of firelight, and they recognized the negro Asa.

"Well, well! Where on earth did you come from?" cried Joe, delighted, and both boys began to crowd the smiling negro with questions.

"I been a-watchin' you boys a good while," said Asa, laughing; "I did n' know who you was till I seen Charley's face over de fire, den I come a-jumpin'. So yuh we all is togedder agin."

"And you got away from the deserters that day, after all?" asked Joe.

"And did you swim across the big prairie like we did?" asked Charley.

"Who, me? I come thoo de woods. I des got away las' night."

Asa's story was, in substance, that after marching him back to the spot where the dog had been killed, the deserters scattered, and lost much time in searching for the trail supposed to have been taken by the two boys. Meanwhile Asa was sent on to camp under the guard of Bud Jones, who left the recapture of the boys to the others. Late in the day the rest returned to the island crestfallen and in great ill-humor.

The negro, after serving as the butt of much violent language and having been threatened with dreadful punishment if he attempted to escape again, was liberated, and allowed to go about his usual employment. In the course of the afternoon and evening, he noted that, while the successful escape of the two boys seemed to cause the other

men great annoyance and dread, Bubber Hardy looked
more cheerful than he had done since he first showed that
he was "hurted" in his mind, after listening to Joe's
memorable speech. Events then took their usual course
in the deserters' camp, and a week passed.

"Yistiddy mornin'," continued Asa, "Mr. Jackson and
four or five de others started off on a trip to meet dey wives
some'rs on de edge o' de swamp, an' I yeared some o' em
say dey did n't calculate to git back for two or tree days.
Well, las' night, wut you reckon, my boss an' de others
went to bed in de loft an' forgot to fasten me up in de
pen, an' soon's I knowed dey was all sleep good, I come
a-kitin', an' yuh I is."

With some assistance, mostly in the form of excited inter-
jections, from Charley, Joe told the story of their adventures since the separation from Asa. A full understanding
arrived at, and the proposition made that the three start
for Trader's Hill at dawn, Asa took up the preparation for
supper where the boys had left off, and they were soon
satisfying their hunger with broiled fish and fowl.

Their meal was not quite finished when the sound of
hurrying feet arrested their attention. Starting up, they
promptly discerned the forms of six men closing in upon
their camp-fire from almost as many different quarters.
Evidently they were to be captured, and every avenue of
escape had been designedly cut off.

"Hit's de 'zerters," whispered Asa, — "Mr. Jackson an'
his crowd. No use to try to run."

They were indeed caught, and it would be useless to
resist. In a few moments the deserters were upon them,

and, seizing the two boys and the negro, they promptly tied their hands.

"So h-yer you is, is you?" cried Sweet Jackson, in scorn and triumph. "Thought you'd git clean out o' the swamp by to-morrow, did you? Well, we'll see about that. Bubber Hardy is willin' to let you boys go, but the balance of us ain't sich natural-born fools. As for this cussed nigger, I don't want to lame him so he can't walk, but jes' wait tell we git him back on the island. *We'll* make him see sights, Bubber Hardy or no Bubber Hardy."

Asa submitted without a murmur, and Joe was for the time so dazed by surprise and chagrin that he could not speak. But Charley began forthwith to cry, and sobbed piteously for half an hour.

"I'll make you sorry for this one of these days," Joe burst out at last, hot indignant tears starting in his eyes.

"You'd better keep a still tongue in your head," rejoined Sweet Jackson, with anger. "That's all I've got to say to you, Mr. Smarty."

Their belongings having been picked up, the prisoners were now led away. A tramp of half a mile brought them into the neighborhood of another camp-fire about which several forms were moving. At a nearer view these proved to be women, four in number.

"I reckon dem's dey wives dey come out yuh to meet," whispered Asa to Joe.

"You caught 'em?" called out one of the women, in a shrill, high voice, as the party approached.

"Yere," was the answer.

It was learned later that one of the deserters, starting

out from the camp with his gun, had discovered the prisoners, and returning, gave the alarm.

" Let me git a look at them boys," said the same woman, as the prisoners were led within the light radiating from the fire.

She was dressed in a coarse homespun frock and sunbonnet, and her sallow face was far from handsome; but she had a bright black intelligent eye, and she gazed at Joe and Charley with great interest, and in a not unfriendly way.

" Oh, Sweet, why can't you turn 'em loose and let 'em go ? " she asked, after staring hard for a few moments. " I know they 're powerful homesick ; I 'm sorry for 'em, — they 're sich putty-lookin' boys."

" They kin make a fool o' Bubber Hardy, but they can't make a fool o' me," was the only answer.

" But what harm kin two little *boys* do you if you *do* turn 'em loose ? "

" They kin tell on us an' git us arrested, — that 's what harm they kin do," answered Bud Jones, dryly.

" You jes' better 'tend to yo' own business, Nancy," said Jackson, gruffly, and the discussion stopped there.

Nancy Jackson — for it was at once clear to the prisoners that she was Sweet's wife — appeared to be the leading spirit among the women. The other three seemed to have much to say to their several lords, with whom they sat apart on the grass, but Mrs. Jackson was the only one who raised her voice in the hearing of the whole camp.

They were all of the illiterate Cracker class, like their husbands, but were women of no little determination, or they would never have ventured into the jaws of the Okefenokee, so to speak, ten or twelve miles from their homes, attended only by two half-grown boys.   The object of their expedition was to meet and spend a couple of days with their husbands, and bring them a small supply of salt, — an article now very scarce in this corner of the Confederacy, as was indeed almost every other article under the sun.

CHAPTER XVIII.

AFTER eating heartily of the supper which the women
had been preparing for them, the six deserters
lighted their pipes; and for about two hours there was
much animated conversation around the camp-fire, the wife
of Sweet Jackson taking a leading part.

Lying passively on the grass beside Asa and Charley,
his hands still bound, Joe gradually became intensely
interested in what this woman was saying.

" I tell you what, people is seein' sights these days," said
she. " Let 'lone salt, some of 'em ain't got a roof to git
under. The backwoods is full o' refugees from Savannah
and Brunswick and St. Mary's and everywhur else. I see
'em go by on the road most eve'y day in wagins and ox
c-yarts and anything they kin git. I've housed loads of
'em sence they been comin', but I has to turn a heap of
'em away.

" One day two or three weeks back, a powerful stuck-up
set come 'long — or their nigger gal was stuck-up for 'em,
they was meek enough theyselves. They come in a ox
c-yart, — a white-headed ole gentleman and his wife and
two young ladies and the nigger gal. I tole 'em they
could stop over night, but the three ladies would have to

go in one room, and the ole man would have to sleep in
the corn-crib; as for the nigger gal, I could n't say what
I could do with her, but I 'lowed to fix her somehow.

"'Very well, madam,' says the ole man, kind o' proud
and stately; 'there is no choice but to stop. You are very
kind, and I will gladly pay your demands.'

"So they lit and come in, and while the ole man and
his lady sot on the piyaza, and the two young ladies walked
up and down in the yard, that nigger wench slipped in to
look at the room. Would you believe it? She stood up thar
in my *comp'ny* room and looked round and turned up her
nose! Then she lent over and felt of the bed and stuck
her nose down to smell of my colored sheets, and says she,
'Hump! missis can't sleep in dis bed!' Well, sir, I was
that mad, I grabbed the broom-stick and run her out on
the piyaza.

"And the ole man and his lady got up and scolded that
gal good, and they says to me, 'You must overlook it,
madam;' and then they went in and took a look at the
room theyselves. and smiled at me kind o' sad like, and
they says, 'This will do very well, madam, and we are
greatly obliged to you.' But, all the same, them three
ladies did n't sleep in that bed, — they slept *on* it. I went
in and took a look next mornin' when they come out to git
somethin' to eat, and I seen how they worked it. They
did n't nair one of 'em git between them sheets; they jes'
spread over the bed a lot o' shawls and things they had in
that c-yart, and laid down on top of 'em. I never *seen*
sich a stuck-up set.

"But I was goin' to tell you 'bout the gibberish them

two young ladies talked when they was walkin' up and down in that yard. I went out to the well to git a bucket o' water, and they passed closte and I heard 'em, and, sir, I could n't understand a single, solitary word! And when I fetched in the water I says to the ole lady, s' I, 'What sort o' gibberish is them two gals a-talkin' out thar in that yard? I ain't never hearn the like.' And she sort o' smiles, and she says, 'I suppose my daughter is speakin' French with her governess, as she ginally does when they 're by theyselves.' Thass jes' what she said. I 's had to deal with a heap o' partic'lar travellers," concluded Nancy Jackson, "but this was the particlares' crowd yit."

The present great scarcity of the necessaries of life, particularly of salt, and the discussion of the subject among the six deserters, gave occasion for another story from the voluble and observant Nancy.

"Why, you can't git none for love nor money these days," she declared. "That salt we brought yistiddy was *give* to us by ole Mr. Richard Macy thar in Trader's Hill. He 's been diggin' up the earth in his smoke-houses and gittin' the salt out — 'extractin'' it, he says. I dunner how he does it, but he does it. Hit 's mighty black and dirty, but hit 's *salt*, and ever'body is glad to git it. He don't sell it off for a big price, like some people would, but he gives it away. He says that salt, cep'n a little for himself, is for the wives and widows of the soldiers. Ole man Macy is powerful sot fernent the deserters — turrible down on 'em, shore 'nough — but he 's a mighty good man.

"I was thar to his place with Liza Wilkinson that time she heard John was killed, and, sir, you ought to 'a' heard

11

him talk! We went to git salt, and found him a-readin' out o' the paper the names o' the killed in the last battle, and when he come to John Wilkinson's name, Liza jes' turned white ez tallow and sot thar dumb. Hit was rale pitiful. And ole man Macy, he says, ' Po' child! The Lord help you!' An' dreckly he got started off like a preacher, and got to praisin' up the soldiers that fell in battle, and to runnin' down deserters plum' turrible — well, sir, hit jes' gimme the cole chills; and to tell you the truth, Sweet, I wished I could see you in the army 'long 'side o' John Wilkinson, even if they did kill you."

"You ijit!" was Jackson's angry interjection.

"Atter while," Nancy continued, "Liza got up and walked out; and when I started home, I found her a-lyin' down in the wire-grass 'side the road, and she laid thar so still I thought for a minute she was dead. I went to her, and I says, ' Git up, Liza, an' less go home; hit's late.' And then she got up an' sot on a log, and I tried to git her to take a dip o' snuff to brace her up; but she would n't, and she looked round at me, and she says, ' Thank God, he want no deserter. I ain't got nothin' to live for now,' she says, ' but I 'm better off 'n some folks. I 'd ruther be the widow of a *soldier* than the wife of a *deserter*,' says she, and then she got up and walked on ez proud ez you please.

"If she had 'a' hit me in the head with a hatchet," declared Nancy Jackson, passionately, tears starting in her eyes, "she could n't 'a' hurt me worse. I don't wish you no harm, Sweet; but God knows I ain't proud o' bein' the wife of a deserter, and I went home that night and had a big cry."

"I did n't know you was sich a fool," was the brutal

rejoinder of Jackson, who had started to his feet and seemed ill at ease.

"Liza's got consumption," concluded Nancy, sadly pensive, "and she won't live long nohow."

"And when she dies, she'll go to heaven, where her brave husband is," burst out Joe, beside himself, his voice shaken with emotion.

"None o' yer —— rantin' now!" exclaimed Jackson, fiercely, giving the prostrate boy a kick.

"You coward! you beast!" cried Joe, starting to his feet.

Jackson leaped toward the boy with uplifted arm, but his wife ran between them and stopped him.

"You sha'n't tech him!" she declared. — "less 'n you knock me down, too. Ain't desertin' the army enough, 'thout jumpin' on a half-grown boy whose hands is tied?" she demanded in great scorn.

Sweet Jackson glared at his rebellious wife in a threatening manner, but hesitated, and after a moment turned on his heel. He felt ashamed, not of his intended assault on the boy, but of having thus been balked by a woman, and that woman his wife, in the presence of his associates; and he gave vent to his rage in the repetition of a number of his favorite oaths.

"Lay down and hush now," said Mrs. Jackson, urgently, to Joe. "You ought to know better than to aggervate him."

The four deserters whose wives were present had each built a "brush-tent" for his own accommodation, and it was in these that the women spent the night in the company of their husbands. The other two deserters, and the two half-grown boys who had accompanied the women, lay down

under the open sky around the fire. Here the prisoners
also passed the night, the latter not only with their hands
still bound, but their feet also, — an additional precaution
which was insisted on by Jackson before he retired into
his brush-tent. Asa slept as soundly as usual; but the two
boys, excited and angered by this fresh indignity, lay awake
and talked in low tones during the greater part of the night.

The morning light found the prisoners stiff and cold, but,
in the case at least of the two youngsters, with spirits still
undaunted. When the bonds holding their feet together
were loosed, the boys and the negro could scarcely stand.

The camp was astir at an early hour, and as soon as
breakfast had been despatched, the four women took leave
of their husbands, and, attended by the two half-grown
lads, departed. The six deserters and their prisoners
moved away in the opposite direction.

"Good-by, Mrs. Jackson! I'll never forget you!"
called out Joe, as the two parties were separating.

"Good-by, Joe," said the deserter's wife, the soul of
kindliness and pity in her voice and looks. "Never you
mind, honey. Don't you fret. You two boys'll git home
safe before long. This sort o' thing can't last always."

"Shet up!" ordered Jackson, but neither his wife nor
the prisoners took any notice of him.

"Won't you please send word to papa and tell him where
we are?" pleaded Charley.

"No; she won't do no sich of a thing!" roared Jackson,
ordering his wife to depart and the prisoners to go forward.

Nancy Jackson looked doubtful, hesitating, pained, as
she listened to the little boy's pathetic entreaty. Without

answering, she turned and walked on, the same expression on her face. The three other women and the two half-grown lads were some distance ahead; but she seemed in no hurry to overtake them, and paused several times to look back. As she did so, the boys could see that there were tears in her eyes.

"She deserved a better husband," Joe remarked to his little brother, as they turned to follow their captors.

The boys were told nothing, but well knew that they were now to be taken back to Deserters' Island. Sweet Jackson marched ahead, followed by two of the men; then came the prisoners, followed by Jones and the two remaining deserters, all advancing in single file. The prisoners' hands were still bound, and the cruel leader of the party swore that he would not allow them to be untied until the island was reached.

Necessarily this caused the march through the jungle to be much more difficult and painful for them than it otherwise would have been. Sometimes, when they stumbled and fell, or when they pushed through dense and thorny thickets, being unable to protect themselves with their arms and hands, they received many painful scratches and blows on the face and head. This was hard to bear, and ere long both Asa and Charley begged that their bonds might be loosed.

Joe made no such request; but at length, toward noon, as they entered a space of open pine barrens, after passing through a dense jungle full of thorny brambles, he rebelled.

"I won't go another step unless you untie my hands!" he cried, throwing himself down on the grass. The boy's face was bleeding in several places from scratches just received.

"Jes' let me git a hold o' him!" cried Jackson, turning back when he saw what had occurred, and cutting a long stout oak switch.

"My hands are tied, and I know you are devil enough to beat me to death," said Joe, with blazing eyes and un-flinching calm, "but I won't budge!"

"Now look-a h-yer, Sweet Jackson, this is gwine a little too fur," interposed Bud Jones. "In time of war some of us has to do despe'rte things, but ther' ain't nothin' to jestify you in beatin' that boy."

"'Tend to yer own business!" cried Jackson. "He's got to mind me or take a whippin'."

"What if you can't make him? I kin tell by his looks he don't aim to budge, beat him ez much ez yer will. He's got the spunk of two or three men like some I know. Besides that, he's got right on his side. Hit ain't right and hit ain't reason to make him go thoo these bushes with his hands tied."

"No, hit ain't," chimed in the other men.

"No sense in it nohow," continued Jones, encouraged by the approval of the others. "How in the dickens kin he git away?"

"I depend I know what I'm a-doin'," rejoined Jackson, angrily. He seemed determined not to yield, and gave utterance to many outrageous oaths before he finally cooled down enough to be willing to a compromise.

"Well," he said at last, "you kin untie the boys, but the nigger's got to stay tied."

"It hurts Asa just as much as it hurts us," declared Joe, with the same unflinching manner, "and unless you untie him too, I won't move — I don't care what you do!"

"Oh, Mas' Joe!" exclaimed Asa, who had heard everything, and who gazed at his champion with an expression of countenance in which wonder and gratitude struggled on equal terms.

This was the occasion of a fresh squabble and further conflict of opinion, emphasized by strong oaths; but in the end the determined boy had his way.

The party reached the deserters' island camp at sundown, and great was the surprise and sensation caused by their arrival. The half-witted Billy was more than well pleased at the return of the two boys, capering around them and shouting in the expression of his delight. But Bubber Hardy became very angry when he learned that Joe and Charley had been captured within ten miles of their home and brought back to the island prison, and he did not hesitate to speak his mind. It was as much as the other men could do to prevent a hand-to-hand encounter between him and the furious Jackson. Even after the boys had been given some supper, and had climbed into the familiar loft and lain down to sleep, they heard the two men still quarrelling over the camp-fire.

"Got to be a deviation somewhere," muttered Asa, as he was shut up for the night. "Ez Mis' Jackson tole de boys, dis sort o' bizness can't last forever."

CHAPTER XIX.

THE PROBLEM IS SOLVED.

A T breakfast the next morning Joe observed that neither Bubber Hardy nor Sweet Jackson seemed disposed to talk. The former looked depressed, the latter sullen; and such conversation as there was had no reference to either, or their recent and violent quarrel. The two leading and conflicting spirits of the camp appeared to have agreed on a truce, or to be biding their time. The boy may be pardoned for hoping that truce there was none, since this would almost inevitably result in the continued detention of the prisoners.

Joe also noted that Lofton's wound was fast healing, but thought it likely that he would wear to his grave an ugly scar all across the left side of his forehead and his left cheek. The covert, unfriendly glances which he now and then directed toward the sullen Jackson were proof to the observant lad that he meditated revenge.

After breakfast Hardy called Joe aside and asked for an account of his and Charley's wanderings since the night of their escape from the island. This the boy very willingly gave, being desirous to please the only friend — barring Asa — whom he and his brother could rely on while in their present position. He, however, spoke guardedly of their experiences on Blackjack Island, being unwilling to let slip the remotest hint of the plans of the distinguished man residing there. The boy was too astute not to have begun long ere this to suspect that his much-named friend had exaggerated his own importance; but he still felt confident that the solitary denizen of Blackjack was, as claimed, a soldier, and that he had designs on the deserters. So he merely stated the fact that they had found a hunter on that island who had been very kind to them.

"Oh, you run up on George Smith, did you?" asked Bubber, smiling.

"Why, do *you* know Mr. George Washington Jefferson Jackson Smith?" asked Joe, amazed.

"Yes, I know him. I reckon he told you a long string o' lies, did n't he? That 's like him. He 's the biggest liar, the biggest coward, and the cussedest fool I ever laid eyes on. He was the first deserter to locate in this h-yer swamp, and he 's been in h-yer gwine on three year."

"What!" gasped Joe, his faith in mankind quaking. "Then he is not the great soldier and counsellor, the friend of General Lee?"

"He never laid eyes on General Lee. That piny-woods Cracker the counsellor of General Lee! He want nothin' but a common foot soldier, and he want that long. He deserted after the first battle he was ever in."

"Well, he fooled *me!*" exclaimed Joe, greatly crest-fallen, and almost ashamed of himself. "But I thought there must be something wrong about that man," the boy declared, after a moment.

"I thought he looked like a ram-goat," said Charley, who

had approached and overheard the greater part of what had been said.

"He told me he was detailed to look after the deserters in this swamp," continued Joe.

"Oh, he did, did he?" laughed Bubber. "Sometimes I think George Smith's brains must be half addled when he gits started on a yarn. *He* a great soldier! Why, he turned and run the very first time he was under fire. They tell me he went runnin' and hollerin', 'Oh, I wish I was a baby! I wish I was a *gal* baby!'"

"At least he has a kind heart," Joe was generous enough to say, after having laughed until the tears ran down his cheeks.

It so happened that the deserters scattered widely that afternoon, and the camp was almost deserted. For some time no one seemed to be left on guard but Sweet Jackson, who lay upon the grass and dozed. Joe watched this man, their worst enemy, narrowly, thoughts of an attempted escape in his mind, as he stood cleaning his gun not far away. Asa worked among his pots and pans at the fire, talking with Charley. The hapless Billy, after being absent for an hour, had shown himself again, and now squatted in the grass just beyond the borders of the clearing.

It was about four o' clock when Sweet roused up and stood erect, calling roughly for some water.

"De ain't none fresh; lemme go git you some fresh," said Asa, hastily, taking up the tin bucket as he spoke.

"Never mind; go on with your work," said Sweet, yawning. "I'll send Billy. Billy is *my* nigger. Billy! Oh, Billy!" he called.

But Billy made no answer. Asa indicated the where-abouts of the boy, and Sweet took a few steps forward.

"You Billy! Why don't you answer me?" he called angrily.

But the boy seemed to be absorbed in contemplating some object on the ground in front of him, and gave no sign of hearing.

"I depend I'll everlas'nly make him hear me!" cried Sweet, enraged, breaking a long stout switch and stripping off the leaves.

The absorbed Billy did not even turn his head when the sound of hurried footsteps in the grass fell on his ear. Not until the switch descended heavily on his back, did he start and look up with the air of one rudely awakened from a dream.

"I'll l'arn you to fool with me!" cried the infuriated Sweet, raining down blows, beneath which the boy seemed to stagger as he attempted to rise. But once upon his feet, he leaped forward beyond reach, and faced his foe, a strange glow in his eyes.

Sweet sprang after him with uplifted switch, when he suddenly became aware that he had trodden upon some soft living body, which yielded beneath his weight and struggled in a peculiar, writhing way. At the same instant he heard a harsh rattling sound, and, as his glance swept downward, he saw that he stood upon a rattlesnake.

Had he kept his position, he might have escaped un-harmed, for his feet were on its body near the neck. The reptile, probably sharing Billy's strange trance, had been, like him, taken unawares. But Sweet in his sudden terror leaped upward and forward. As he moved, the rattler struck him on the right leg just above the ankle. The effect of the man's leap was only to fasten securely in his flesh the snake's hooked fangs. Uttering wild cries, the unfortunate deserter dashed hither and thither, dragging after him the struggling snake.

A laugh at such a moment was truly the most unexpected and cruel thing in the world, yet that is what Joe, Charley, and Asa, who had drawn near, now heard. They knew without looking that it was the half-witted boy who laughed. He did not stop there; he danced about, and shouted again and again, —

"That's right, son! Stick to him, son!"

Charley knew then that the snake was the pet which he had once been permitted to see.

"That's right, son!" shouted Billy. "Give it to him! That's what he gets for jumpin' on me."

Calling madly for help, Sweet ran staggering toward the camp.

"Beat him off o' me! Beat him off o' me!" he cried, looking toward Asa and the boys.

The rattler was as much a prisoner as his victim, and would gladly have let go and escaped. Had Sweet seized the snake by the neck and lifted it, the fangs could have been loosened in a moment; but fear seemed to deprive him of reason, and he did nothing but spring about and yell.

"We must do something," cried Joe, recovering from the stupefaction of the first few moments. Seizing an axe, he ran forward and dealt the snake a blow, severing a few inches of its tail, but not loosening its unwilling hold.

Immediately after this, Sweet stumbled and fell prone on the ground, crying out the more from fear of closer contact with the snake. But the effect of the fall was to loosen the imprisoned fangs, and the rattler would now have glided rapidly away, had not Joe and Asa set upon it with sticks, quickly despatching it, much to the indignation and sorrow of Billy.

This done, they turned to the unfortunate Sweet, who was now tearing off shoe and sock in a hurried, terrified way, and groaning aloud. The wound had already begun to swell.

" Can we do anything for you, Mr. Jackson ? " asked Joe.

" Oh, I don't know what to do ! " was the despairing answer. " Run, go call Bubber and the rest of 'em. Maybe they'll know."

Joe and Charley then ran out of the clearing, shouting, and in about twenty minutes returned with Bubber and three of the other men. As they approached they saw Asa preparing to cut the body of a fresh-killed partridge in half, the neck having just been wrung off. Sweet now lay upon his back on the grass, shuddering with horror.

" If anybody's got any whiskey hid off anywhere," said Bubber, in a tone of authority, " let's have it. Now's the time to fetch it out."

He looked from one face to another, as heads were shaken, until one of the deserters turned and moved away, remarking that he had a "leetle smodgykin" saved up for a time of need, and would get it. He walked off into the woods, and returned shortly with a small bottle containing less than half a pint of colorless whiskey. This was forthwith poured down Sweet's throat.

Stout cords were then tied as tightly as possible round the leg above and below the wound, in order to check the circulation of the poisoned blood, and the raw quivering flesh of the partridge was pressed hard on the wound itself, acting as an absorbent.

Several birds were slain, one after another, and as soon as one bleeding half was taken from the wound another half was applied. Asa had suggested that the raw flesh of the rattler be applied in lieu of the partridge; but this the poisoned man would not permit.

But by nightfall Sweet's leg was startlingly swollen, and he had begun to wander in his mind. It was plain that too much time had been lost while the snake hung from its victim, and while the men were being summoned.

Charley had meanwhile described how he had one day been invited to visit the snake at its hole; how Billy had fed it, and seemed to be on friendly and familiar terms with it. Joe and Asa also testified that the boy, having evidently enticed the snake to the clearing, was playing with the reptile when Sweet set upon him with the switch. No one forgot that Jackson was of an ugly temper, and treated the poor boy cruelly; but none the less was Billy now looked upon with suspicion and aversion, and by common consent he was shut up in the prison-pen built for Asa.

The majority of the men seemed to suspect that he was no less than a fully equipped conjurer; and the next day some of them took the precaution of putting red pepper in their shoes as a safeguard against witchcraft. The poisoned man grew worse and worse; and soon after midnight he died in great agony.

After this a profound hush fell on the bustling camp. Joe and Charley retired to the loft; but all the men sat about the fire and watched till break of day. Arranging the limbs and covering the face of the dead, they freshened the fire and sat down to wait with wide-open eyes and busy thoughts. Their vigil was not merely to protect all that was left of Sweet from the possible attacks of wild animals, but to conform to the custom of their people. Moreover, no one cared to sleep. Men who had scarcely reflected in their lives felt impelled to do so now. Each thought upon past deeds and upon future amends.

The blow that had fallen seemed to them not merely a judgment on their dead friend, but on them all, because of the selfish and unlawful life which they were living. But when at last the morning broke, only one of the eight still kept faith with his resolves of the night. The others had felt no more than that sham repentance which is active only when in the presence of fear.

Awaking rather late next morning, Joe and Charley heard the sound of carpenters' tools, and, descending the ladder, saw several of the men engaged in making a rough coffin. Others were digging a grave several hundred yards out on the open ridge. By the time Asa had given the boys something to eat, the coffin was ready and the body was placed in it. Then four of the men lifted it, and bore it to the grave, followed by all except Billy, who was still in prison.

One of the deserters, called Arch Thatcher, had formerly been a lay preacher. He now offered a prayer, sung a hymn, in which a few others joined, and made a few remarks about the vanities of the world, after which the coffin was lowered and the earth thrown in. It was then, as all were ready to return to camp, that Bubber cleared his throat and stepped forward.

"I don't know hardly what to say, men," he began, paused, then continued: "I don't know how it is with you-all; but as for me I don't feel right, and I aim to make a change. I'm tired playin' sneakin' suck-egg dog, and from this on I expects to try to be a *man*. I'm a-goin' back to the fight myself; I don't care what the rest of ye do. You kin stay right on h-yer, men, if you hanker to stay, and I won't tell on ye; but as for me, I'm a-goin' to take these boys home and then go back to the fight. Anybody got anything to say agin it?"

He paused and looked around. No one spoke. Joe's, Charley's, and Asa's were the only bright faces which met his gaze. The others were downcast.

"I got just one thing to ask o' you-all," he continued, looking at one or two of the leading spirits among the men. "I want to ask you to take Billy home to his people. You know whar to find 'em. 'T ain't fur. Sweet was kin to Billy himself, but I'm free to say he didn't have no right to fetch him in h-yer."

"We'll be mighty willin', I'm a-thinkin'," responded the man called Thatcher. "We'll be glad enough to git rid of 'im. We don't want no sich around. Fust thing we know he'll be tolin' up another rattlesnake."

"I'm a-goin' to take Asa and the two boys and start to-day," announced Bubber. "And I'm a-goin' to take my share of the skins, too. We'll have to take two o' the boats; but we'll leave 'em in the old place on t' other side the prairie, and to-morrow three of ye kin go over in t' other boat, and bring 'em all back. Now if anybody's got anything to say agin it, let him say so right now, and we'll settle it right h-yer 'fore we quit."

But no one made a reply, and the plan was understood as settled. Dislike the arrangement as they might, none of the men felt disposed to stand forth and challenge the "cock of the walk."

Calling Asa, Bubber ordered him to proceed at once to the cooking of a "snack" for their proposed journey, then turned away to make other preparations on his own account. Left to themselves, the two happy boys were not slow to collect their few treasures and otherwise prepare for the march.

The other seven men hung about the grave, talking gloomily and in low tones, not, however, of the virtues or vices of the dead, but of their own situation and doubtful prospects. The dead man had few, if any, real friends, having maintained the ascendency which he enjoyed, not by the power of sterling character, but by the force of will and muscle. His truculent nature had often been the subject of comment with the two boys; but since the hour of the tragedy their thoughts had been filled only with pity for the unhappy man who was less their enemy than he was his own. As they turned away after the burial, however, Charley gave expression to a thought which was in Joe's mind also.

"Well," said the little fellow, innocently, "I hope Mrs. Jackson will get a better husband now."

12

CHAPTER XX.

HOME AT LAST.

TWO hours later the seven deserters saw the last of their former comrade, as the boats pulled away from the landing, and began the difficult struggle across the prairie. One carried Bubber Hardy, together with his hides, strapped in two small, but heavy bales. The other contained Asa, Joe, and Charley, who was the last to step on board, he having halted in order to peep through the cracks of the prison-house and call out a good-by to his hapless friend Billy.

It was five o'clock in the afternoon when they landed. Saying they had no time to lose, Bubber gave one bale of hides to Asa, shouldered the other himself, and led the way. The boys followed with their own belongings.

After a march of some three-quarters of a mile through a forest, which thickened as they proceeded, the skins were thrown down under a tree and abandoned; and the party pushed on a full half a mile farther before a spot suitable for a camp was found.

The sun was not yet down; but it was by no means a waste of time to halt. Bark must be stripped from the cypresses to spread on the damp ground; moss or leaves must be gathered, in order to soften what would otherwise be a very hard couch ; fuel must be collected, a fire built, and supper cooked. Giving Asa some directions, Bubber walked off into the woods. An hour later it had grown dark, and he had not yet returned.

"He must have changed his mind, and gone back to the deserters," said Joe at last, uneasily.

"He des gone off ter hide dem skins, — dat wut he up ter," was Asa's confident rejoinder ; and a few minutes later Bubber reappeared.

It seemed to the boys that their hardships and miseries were already over. They ate heartily of the supper, slept soundly all night, and during the long, difficult march of eight hours next day, did not once straggle behind or lose heart. When they finally entered the open pine woods beyond the limits of the swamp, they could scarcely restrain shouts of delight.

Joe particularly felt happy. His great plan had indeed failed; but still his hopes were in a measure realized. He had not persuaded a whole band of deserters to return to the war; but after a long sojourn among them in the fastnesses of the Okefenokee, he was now on the threshold of the outer world, accompanied by the lost Asa and at least one penitent, convinced of the error of his ways.

As the familiar double-pen log-house came into view, the boys were gladdened at sight of smoke issuing from the chimneys. Somebody was there; perhaps their father and mother and sister. They quickened their steps, looking forward expectantly.

Drawing nearer, they observed with surprise that a soldier stood all alone at the gate. He saw them almost at the same moment, and, after a searching glance, he walked hurriedly to meet them.   It was Captain Marshall.

"Joe! Charley! Is it possible?" he exclaimed, when they had met, putting his hands on their shoulders in a glad way.   "Where have you been?   The whole country has been searched for you."

In a few hurried words Joe outlined the story of their adventures, not forgetting to mention Bubber's resolve to re-enlist.

"I reckon they won't shoot me if I give up and go back to the fight, will they, Cap'n Marshall?" asked Bubber, humbly, with the air of one prepared to meet his fate.

"Your repentance comes too late," answered the captain, sternly and sadly.   "The war is over."

The deserter started as if he had received a blow, and drew back, his face a living picture of shame and regret.

"And we brought Asa, too," cried Joe, proudly, not taking in the captain's meaning, so great was his joy and so turbulent his thoughts.

"Too late again," said Captain Marshall.   "Asa is now free."

Asa looked about him in bewilderment, and Bubber repeated mournfully, "The war is over!"

Joe caught the words this time, and, with a great gulp in his throat, asked what all this meant.

"General Lee surrendered at Appomattox on the ninth of April," replied the captain.

"And I can never be a soldier!" exclaimed the boy, in great sorrow, after asking a few more questions.

"You can at least be a brave man," said Captain

Marshall, no less sadly. "But run into the house, boys; be quick!" he added, turning to move away. "Go to your mother and sister. They have been almost distracted about you."

After their mother and sister had kissed them many times and wept over them; after their dear old father had held them against his heart, and all had looked at them long and fondly; after many questions had been asked and answered, and their long story had been told almost in detail; after night had fallen, and the reunited family were seated together over their evening meal, — Joe remembered the partner in their late misfortunes, and abruptly addressed his father, making an unexpected request.

"I want to give my gun to Asa," he said. "He has none, and I know he wants one. May I, father?"

"Yes. He deserves to be rewarded."

"I want to give him something, too, papa," cried Charley. "He was *so* good to us! You ought to have seen him when we were runnin' from the deserters. He let me ride on his back a long ways, and I know he was tired. What can I give him, papa? I could give him my hatchet, but that would n't be much."

"I'll tell you what you can give him," said the father, well pleased to see these generous impulses in his sons; "you can give him a piece of land. He is free now, and may want to set up for himself. I am not a rich man any longer, but I can afford to give Asa a few acres. I'll give *you* the land, and then you can give it to him."

"Oh thank you, papa," cried Charley, delighted, and soon ran away to tell the negro of his good fortune.

"Mr. Hardy was very good to us, too," said Joe. "But for him, we should have had a hard time in that deserter camp. I hope Captain Marshall and the soldiers won't do anything to him."

"He will not be molested now. The war is over, and the remnants of our armies are disbanding everywhere. But he will be disgraced for life, and deserves his fate, however kind he may have been to you.",

Though a sense of strict justice might dictate it, to the boy this speech seemed stern, considering the deserter's active and complete repentance; and he could not help hoping that Bubber Hardy would in time win the full confidence and respect of his fellow-men.

His father told him that night that it was well the war was over; but Joe was a long time in recovering from his first feeling of disappointment and regret. Not so Charley, who became deeply absorbed in other things as soon as Martha confided to him a great secret, — which was that she was engaged in baking a wedding cake.

THE END.